Forensic Interpretation of Glass Evidence

Forensic Interpretation of Glass Evidence

James Michael Curran, Ph.D.
Tacha Natalie Hicks, Ph.D.
John S. Buckleton, Ph.D.

with contributions by
José R. Almirall • Ian W. Evett • James A. Lambert

Boca Raton London New York Singapore

A CRC title, part of the Taylor & Francis imprint, a member of the Taylor & Francis Group, the academic division of T&F Informa plc.

Library of Congress Cataloging-in-Publication Data

Curran, James Michael.
Forensic interpretation of glass evidence / by James Michael Curran, Tacha Natalie Hicks, John S. Buckleton.
p. cm.
Includes bibliographical references and index.
ISBN 0-8493-0069-X (alk.)
1. Glass. 2. Forensic engineering. 3. Ceramic materials. 4. Bayesian statistical decision theory. I. Hicks, Tacha Natalie. II. Buckleton, John S. III. Title.

TA450.C87 2000
616′.1—dc21 00-030354

Direct all inquiries to CRC Press, 2000 N.W. Corporate Blvd., Boca Raton, Florida 33431.

International Standard Book Number 0-8493-0069-X
Library of Congress Card Number 00-030354
3 4 5 6 7 8 9 0
Printed on acid-free paper

Preface

In 1933 the problem of identifying certain minute splinters of glass was referred to the Dominion Laboratory in New Zealand. The splinters from the corner of an attaché case of the accused were alleged to be the result of the case being used to break a shop window preparatory to taking goods.[1] Refractive index and density were used in the comparison. Of 65 samples previously encountered only 1 matched in all respects to the glass from the attaché case and that 1 was the plate glass from the shop window.

The analysis of glass evidence for forensic uses was an exciting topic even before Nelson and Revell[2] first carried out their backward fragmentation experiments and Ojena and De Forest[3] perfected the method of characterizing a fragment of glass by its refractive index.

This field of glass evidence interpretation was revolutionized in the late 1970s when a young Forensic Science Service (FSS) document examiner named Ian Evett[4] decided to introduce statistics as a method for consistent and objective evaluation of forensic glass evidence.

Evett discussed the problems facing forensic scientists with Professor Dennis Lindley. Lindley can be considered as one of the forefathers of modern Bayesian thinking, given that he regularly entered into debate over the validity of the subject with the great Sir Ronald Aylmer Fisher who is considered the founder of modern statistics. At the same time that Dr. Evett introduced statistics into the analysis of glass evidence, Professor Lindley remarked that the actual solution was a Bayesian one.[5]

This book evolved from an "interpretation manual" written by Evett, Lambert, and Buckleton for the FSS. Some material has been added and other areas updated. We also acknowledge Dr. José Almirall for the substantial amount of material he contributed to Chapter 1.

We have followed the Bayesian method of reasoning in this book. This has been a deliberate choice, and we have not argued its advantages extensively here. It may become apparent to anyone reading this book what these advantages are simply by noticing that the Bayesian approach has allowed us to handle some difficult casework problems with the assistance of logic. It would complete the picture to give a critique of the frequentist approach. This is surprisingly hard because the frequentist approach is not firmly grounded in logic. No person has offered any coherent and comprehensive system by which the frequentist approach may be applied to highly variable

D.F. Nelson and B.C. Revell, Backward fragmentation from breaking glass, *J. Forensic Sci. Soc.*, 7, 58, 1967 (Reprinted with the kind permission of the Forensic Science Society).

casework situations, or how this frequency applies to the questions before the court.

By using either method, most practitioners can interpret simple evidence such as one bloodstain left at a scene that matches a suspect. However, misleading statements and ad hoc solutions may result when the frequentist approach is applied to more complex cases. For instance, we have shown coherent methods to understand the value of the presence of glass per se. We have seen frequentists attempt to formulate this assessment, and both they and we feel that there is great value in this. Furthermore, both the Bayesian and frequentist schools would accept that the larger the group of glass, the higher the value of the evidence (in most cases). The difference is that the Bayesian school can make a logical attempt to evaluate this evidence.

It is possible to expose these differences with a simple case example. Consider a case where a man has been seen to shoulder charge six windows. A suspect is apprehended 30 minutes later and found to have two fragments of glass upon his highly retentive black jersey. These two fragments match one control. Simply quoting the frequency with which this match would occur, say 1%, might imply some evidence supporting the prosecution hypothesis. Presented with this case, frequentists perform one of several actions. The more thoughtful of them start to say something along the lines that this does not appear to be much glass given the circumstance. In doing this, they are unconsciously evolving toward the Bayesian thought.

When computing power and the statistical tools required became available in the early 1980s, Ian Evett, with the help of John Buckleton, Jim Lambert, Colin Aitken, and Richard Pinchin, developed an approximate Bayesian solution which was implemented and used in forensic casework.[6,7] We call this implementation approximate for two reasons: (1) a full Bayesian treatment requires the evaluation of the entire joint distribution of the control and recovered samples (a task which still defies solution today), and (2) the approach still contains a "match"/"nonmatch" step. This second point was remedied by Walsh et al.[8] in the mid 1990s.

Since then there have been rapid advances in the statistical analysis of forensic glass evidence and the statistical evaluation of many other types of forensic evidence, the most notable of these being DNA. In the U.S. and many other countries around the world, forensic scientists have found the major focus of their work shifting toward the evaluation of DNA evidence. This initially involved the analysis of nuclear DNA found in blood, semen, and saliva, but in recent years mitochondrial DNA has also become an important source of forensic information. However, law enforcement is beginning to realize that DNA is not always available and there is a whole field of forensic science called trace evidence analysis.

It is for this reason that the authors decided to write a book on the statistical interpretation of glass evidence. Glass work accounts for some 20% of casework in New Zealand and 12% in the U.K. In the past, the U.S. has not had a strong history of forensic glass analysis, but with large amounts of federal funds formerly devoted to military applications being set aside for forensic research, we expect this to change rapidly over the next few years. The U.S. is at the forefront of research in the use of elemental concentration data as a means of discriminating between different glass sources.

This book is intended for forensic scientists and students of forensic science. There have been enough treatises written by statisticians for statisticians. Of the six contributors, five are forensic scientists, one is a statistician, and all have done physical glass analysis and presentations of statistical evidence in a court of law. The intention of this book is to provide the practicing forensic scientist with the necessary statistical tools and methodology to introduce statistical analysis of forensic glass evidence into his or her lab. To that end, in conjunction with this book, we offer free software (available by E-mailing James Curran at: curran@stats.waikato.ac.nz.html, courtesy of James Curran and ESR) that implements nearly all of the methodology in this book. We offer this software with the minor proviso that users must supply their own data because we are unable to give away some of the data sets mentioned in this book.

The examples and theory in this book primarily revolve around refractive index measurements. However, where applicable, the methods have been extended for elemental concentration data. Caseworkers who deal with mostly elemental analysis should not be discouraged by this apparent slant. The theory is easily transferred to the elemental perspective in most cases by simply substituting the equivalent elemental measurement.

Chapter 1 is an introduction to the physical properties and methods for analysis of forensic glass analysis. This chapter is intended for forensic scientists new to the area of glass analysis, students of forensic science, and perhaps statisticians or lawyers who are interested in the physical processes behind the data.

Chapter 2 provides an introduction and review of the conventional or classical approaches to the statistical treatment of forensic glass evidence. Topics covered include range tests, hypothesis tests, and confidence intervals. Grouping of glass considered to have come from multiple sources, *t*-testing, coincidence probabilities, and a simple extension of the classical approach to the analysis of elemental data. By the end of Chapter 2, the reader should be able to perform a two-sample *t*-test for the difference between two means, carry out a range test, construct a confidence interval, and have a basic understanding of the statistical procedures necessary for automatic grouping of recovered glass samples.

Chapter 3 offers the reader an introduction to the application of Bayesian statistics to forensic science. Bayesianism is an entirely different approach to the subject of statistics. A Bayesian approach often requires more thought about the problem involved and the results desired. In this chapter the reader will learn the reasoning behind the Bayesian methods and hopefully gain some insight as to why this approach is preferred. In addition to Bayesian thinking, Chapter 3 introduces the rules of probability, the details of a Bayesian approach to the statistical analysis of forensic glass evidence, and many worked examples to aid comprehension.

In Chapter 4 we attempt to summarize the experimental knowledge gained to date in the fields of glass frequency surveys and the prevalence of glass on clothing.

Chapter 5 describes survey work on transfer and persistence of glass. Many of these works will be familiar to the experienced glass examiner; however, they are presented with the deliberate attempt to make them relevant in the Bayesian framework and with some novel comparative work.

Chapter 6 discusses the particular statistical tools and data that are necessary for a Bayesian approach to be implemented. In particular, the reader is introduced to histograms and their more robust extension and density estimates. This chapter also discusses the various software packages available for the evaluation of glass evidence.

Chapter 7 covers the difficult task of reporting statistical information in statements and *viva voce* evidence. It covers the "fallacy of the transposed conditional," an error that has led to appeals in the field of DNA evidence.

The Authors

James M. Curran, Ph.D., is a statistics lecturer at the University of Waikato, Hamilton, New Zealand, with extensive publications in the statistical analysis of forensic evidence. In 1994 he was awarded a scholarship from the Institute of Environmental Science Ltd. (ESR) to work on statistical problems in the analysis of forensic glass evidence. He received his Ph.D. from the University of Auckland, New Zealand, in 1997. During this period, Dr. Curran lectured and gave seminars at numerous conferences and universities in New Zealand and overseas. He also developed software that is now used for day-to-day casework in New Zealand and as a research tool in several laboratories around the world.

In 1997 Dr. Curran was awarded a postdoctoral fellowship from a New Zealand government agency, the Foundation for Research in Science and Technology (FORST). This provided Dr. Curran with funding to go to North Carolina State University in Raleigh, NC, for 2 years and work on statistical problems in DNA evidence. While in the U.S., Dr. Curran provided statistical reports or expert testimony in nearly 20 criminal cases involving DNA evidence.

The California Association of Criminalists and the U.K. Forensic Science Service awarded Dr. Curran the Joint Presidents Award for significant contribution to the field of forensic science by a young practitioner.

Tacha Hicks, Ph.D., graduated with honors in forensic science from the Institut de Police Scientifique et de Criminologie (IPSC) at the University of Lausanne. Her doctoral dissertation entitled "The Interpretation of Glass Evidence" explores many areas described in this book such as transfer and persistence, glass found at random, and knowledge-based systems.

Since 1996 Dr. Hicks has been involved in both the European Glass Group (EGG) and the Scientific Working Group on Materials (SWGMAT). From 1993 to 1999 Dr. Hicks worked as a research assistant at the IPSC, supervising student research in glass evidence.

John S. Buckleton, Ph.D., is a forensic scientist working in New Zealand. He received his Ph.D. in chemistry from the University of Auckland, New Zealand. Dr. Buckleton has coauthored more than 60 publications in the field of forensic science and has taught this subject internationally. Dr. Buckleton has helped develop a computerized expert system for the interpretation of glass evidence.

In 1992 Dr. Buckleton was awarded the P.W. Allen Award for the best paper in the *Journal of the Forensic Science Society.*

Contents

chapter one

Examination of glass

José Almirall, John Buckleton, James Curran, and Tacha Hicks

A forensic scientist may be asked to examine glass to reconstruct events (for example, to determine whether a pane of glass was broken from the inside or outside) or to associate a person or an object with the scene of a crime or a victim. The strength of the opinion of the forensic scientist depends on many factors, including the nature of the background information and comparisons used, the type of glass involved, whether it is rare or common, and how well it is characterized.

This chapter will consider the examinations and comparisons that are often used in the evaluation of glass as evidence. This is preparatory to introducing modern and classical methods for the interpretation of the allegedly transferred material. We also include a brief description of glass manufacturing processes and composition of glass, as these are important in understanding the examinations to be performed and the sources of variation to be expected. We cover measurements such as comparison of thickness, color, density, and more extensively the measurement of refractive index and elemental analysis, two of the dominant techniques in modern laboratories. It is our intention, however, to introduce these methods rather than give an extensive review of them.

1.1 History

Glass is defined as "an inorganic product of fusion that has cooled to a rigid condition without crystallization."

The first glass objects used by man probably originated from naturally occurring sources such as obsidian, from which sharp tools may be chipped. It is believed that manmade glass originated in the regions now known as Egypt and Iraq around 3500 years ago.[9] Pliny, the Roman historian, recounts a story whereby Phoenician sailors propped a cooking pot on some blocks of natron (an alkali). They noticed that the sand beneath the fire had melted and assumed the properties of a liquid. Upon cooling, they also noticed that the liquid hardened into the material now known as glass. Modern scholarship suggests that glass developed from faience, an older material made from crushed quartz and alkali. These same ingredients in different proportions form a true glass.

1.2 Flat glass

The development of flat glass manufacturing methods progressed through the early 20th century, first in Belgium with the Fourcault process.[10] In this process, the components are first melted in a large tank (which little by little dissolves itself in the molten glass, so that traces of the tank [zirconium, for example] are present in the glass product). The sheet of glass is then drawn vertically through a slotted refractory in a continuous ribbon. The problem of "necking down" to a narrow ribbon was solved by chilling the ribbon against paired rolls that gripped the sheet edges. Glass made this way has a fire finish or polish which is the brilliant surface achieved by allowing molten glass to cool to rigidity without contact. The solidified ribbon, however, has a noticeable amount of waviness. Figure 1.1 shows a diagrammatic sketch of the Fourcault process.

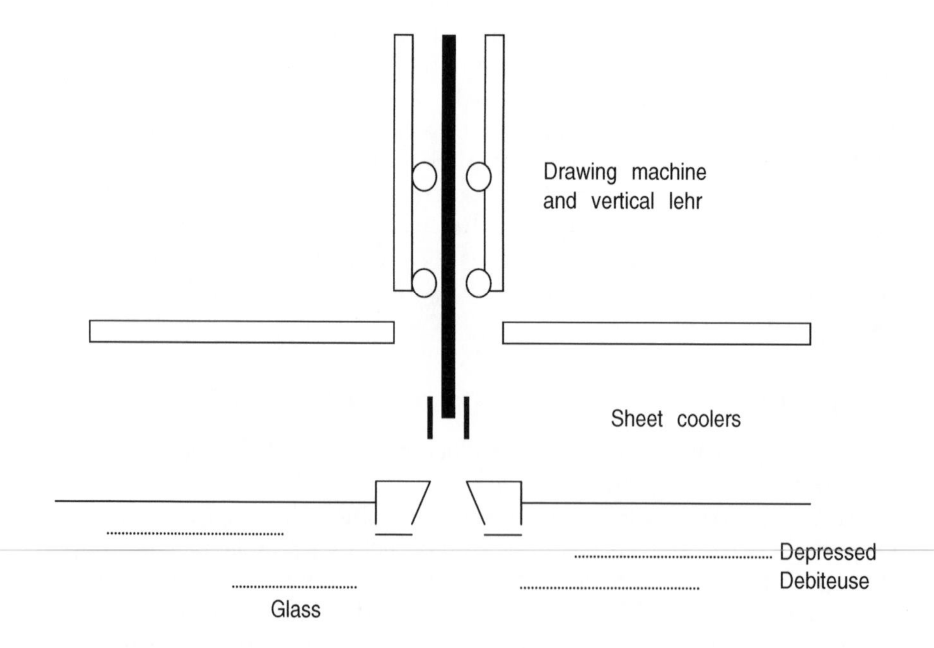

Figure 1.1 Diagrammatic sketch of the Fourcault process.

Later in North America the Libbey-Owens and Pittsburgh methods refined this. The latter replaces the depressed debiteuse with a submerged drawbar, but otherwise the principle and outcome are the same.

Glass drawn from a molten source in this manner is referred to as sheet glass and typically has a fine finish with optical distortions due to minor thickness variations. The term "sheet glass" has been used in forensic texts to mean either any flat glass or glass made by this wire drawn method. This latter definition seems better and will be used in this book.

Sheet glass has a minor amount of visual distortion due to minor variations in thickness. Plate glass was developed to overcome the problem of visual distortion. Typical plate glass production has two predominant steps: first, rough blank production and, second, the grinding and polishing stage. Plate glass was possibly first produced by pouring onto a flat base and rolling to produce the rough blank. A factory in St. Gobain, France was built for this purpose in 1688. Modern machines produce the rough blank by rolling glass directly from the spout and automate grinding and polishing.

Patterned glass may be produced by a number of methods, but mass produced patterned glass is typically made by squeezing the flow of molten glass between paired water-cooled rollers. One or both rollers impress the desired pattern. Wired glass can also be made by feeding wire mesh between the forming rollers.

1.3 Float glass

The float process was developed by the Pilkington company in Britain. The incorporation of a liquid tin bed in the float process leads to a smooth and flat surface. The molten glass is delivered onto a bed of liquid tin where the glass "floats" over the metal. Rollers on the sides pull the glass to a desired thickness depending on the speed of the pull. The float process is currently used for the manufacture of the vast majority of flat glass (see Figure 1.2). It is important to note that the surface in contact with the tin will show luminescence when excited at 254 nm, enabling the forensic scientist to identify float glass (elemental analysis with the scanning electron microscope [SEM] also allows the identification of float glass). It has also been shown that this manufacturing process induces a gradient in the refractive index of the glass, which becomes anisotropic.

1.4 Toughened glass

Flat glass may be processed by tempering. This method is also termed "toughening" and indeed does produce a product that is harder to break. The product is sometimes called safety glass, which reflects the way that it breaks into "diced" pieces that do not have sharp or pointed edges. Tempering is achieved by rapid cooling of the heated glass, typically with air

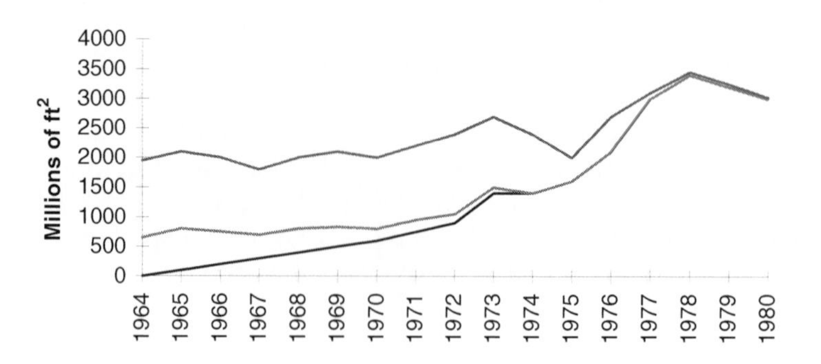

Figure 1.2 Flat glass production in the U.S. from 1964 to 1980.

jets on both sides. This compresses the surfaces and places the center in tension. The process is believed to strengthen (toughen) the glass because surface flaws are thought to initiate breakage. Flaws initiate breakage when the glass is in tension. Inducing surface compression reduces or eliminates this tendency. The center, in tension, cannot possess surface flaws. Modern trends in tempering include the move to thinner glass (a cost-saving measure) and quick sag bending, a shaping technique.

Large fragments of tempered glass may appear in casework in hit and run incidents or more rarely in the pockets of clothing. Such large fragments may be recognized as tempered by their diced appearance and by the presence of a frosted line through the middle of the cube (see Figure 1.3).

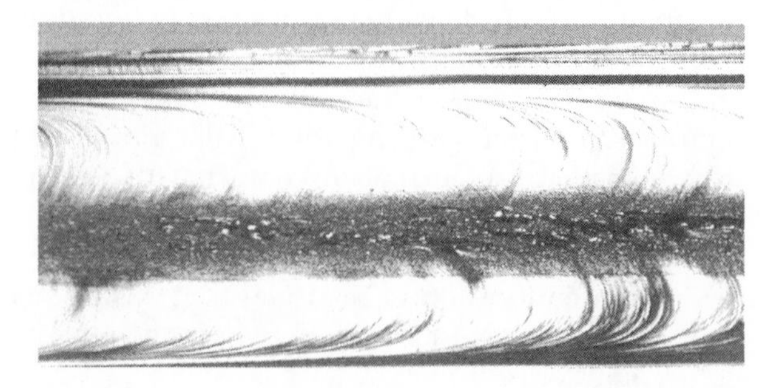

Figure 1.3 Characteristics of tempered glass cubes, such as frosting in the middle of the cube.

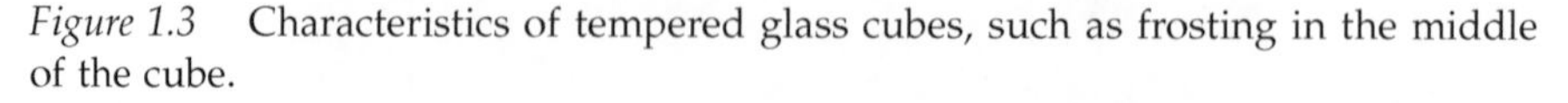

The identification of glass as "toughened" by polarized light microscopy has been reported for fragments measuring more than 20 mm^3.[11] Sanger concluded that correctly mounted and lit fragments exhibiting brightness, if not colors, must be toughened. Small fragments not exhibiting brightness may still be toughened, but below the size required to produce the effect. The authors are unaware of any laboratory using this method. With the small fragments typical of glass recovered from clothing, these observations cannot be made, and for fragments of this size annealing has become the method of choice.

1.5 Laminated glass

Another modern trend is the increased use of laminated glass. This is a composite typically of a sheet of plastic between two sheets of flat glass. Flat or shaped laminates may be produced and are dominant in modern windscreens. The two sheets may be taken from one pallet, in which case they are typically very similar, or from two pallets, in which case they may differ. Some applications deliberately call for color differences in the two sheets. Laminates are noted for their ability to resist penetration, acting as security glass or to restrain passengers within a vehicle during collision.

1.6 Glass composition

Soda-lime glass accounts for the majority of the glass manufactured around the world, producing windows (flat glass) for buildings and automobiles and containers of all types. The major components in the raw materials for this type of glass are sand (SiO_2, 63 to 74%), soda ash (Na_2CO_3, 12 to 16%), limestone (CaO, 7 to 14%), and cullet. The major component (or primary former) is sand, and if it is of very high quality then it is sufficient to produce glass. However, because of the high purity and temperature required to melt sand, soda ash and potassium oxides are added to lower the melting point. Limestone is then used so that the glass becomes insoluble in water and acquires a higher chemical durability. As the products used are not pure, other processes designed to improve purity may be necessary. Of particular concern for the manufacturer is iron or chromium contamination in the sand deposits, which can lead to undesirable coloring.* The impurities in the main products, such as those in cullet, the potassium oxides in soda ash, the magnesium as well as aluminum oxides in the lime, or strontium in dolomite (a source of CaO and MgO), can be used by the forensic scientist to discriminate glass of the same type.

Depending on the end use of the product and on the manufacturing process, special components will be intentionally used.

- Boron oxide (B_2O_3), for example, will be added to improve heat durability in glass intended for cookware, laboratory glassware, or automobile headlamps.
- Lead oxide (PbO) will improve the sparkling effect of glass or absorb radiation (it is also used for stemware, neon, electrical connections, thermometers, and eyeglasses).

* The manufacturer will add CeO_2 (As_2O_3 was formerly used), SbO, $NaNO_3$, $BaNO_3$, K_2SO_4, or $BaSO_4$ to decolorize the glass.

- Other examples are the addition of silver (Ag) in sunglasses and of strontium in television screens in order to absorb radiation.

Depending on its use, glass will also contain different quantities of the same elements. For example, container glass will have a low level of magnesium, iron, and sodium compared to flat glass.

1.7 Glass breakage under impact

In order to understand transfer mechanism, we will discuss breakage under impact. There is a separate body of knowledge relating to breakage under heat stress. As a preliminary rule, glass is strong in compression, but weak in tension. This fact in itself explains a lot regarding glass breakage. However, there are two different ways in which glass may fail under impact.[12] Failure to recognize these two different mechanisms has led to some erroneous thinking. For instance, a toughened "glass will arrest the fall from a couple of stories height of a massive steel ball...but it will collapse like a bubble from the impact of a little steel dart dropped from the height of four inches." This example defines the two fracture mechanisms: the flexural type or the bump check (percussion cone) type.

Despite their names, both types can be produced under static conditions or under impact. Glass breaking under flexion will typically have a break origin opposite the point of stress (or impact) (see Figure 1.4).

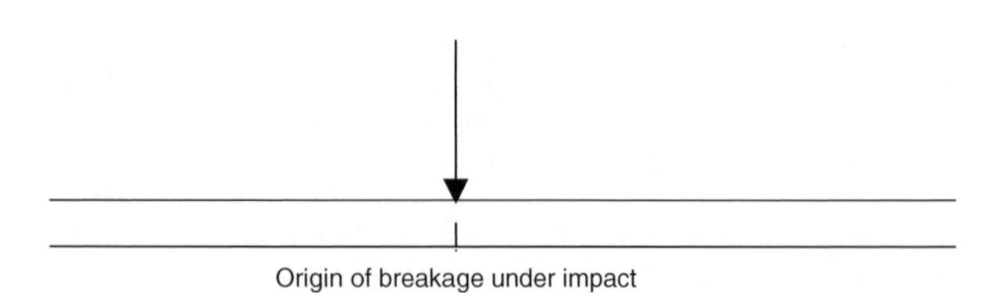

Figure 1.4 The origin of breakage.

The effect of "flaws" as points of initiation has long been known.[13] Flaws, especially on the surface opposite to the force, greatly reduce the strength of the glass to flexion breakage.

Flexural stresses are essentially the same for any object and are determined more by the kinetic energy of the striking object rather than by the nature of the object itself. Therefore, the stress imparted by a wooden bat and a steel ball is similar with respect to flexural stress. Bearing stress and the tendency to produce percussion cones, however, is very different for these two objects.

Key elements, therefore, in examining breakage are flexibility vs. strength of the glass and the nature of the impacting object. An object cannot start a percussion flaw if it is too soft to do so.

Small, hard objects moving rapidly do damage mainly by starting percussion cones (however, refer to the discussion of rifle or gun shots later), whereas massive objects (of any hardness) traveling slowly initiate breakage in flexion.

1.7.1 Breakage in flexion

The concept of breakage in flexion is essentially the same in flat glass and container glass except that in container glass the principle is modified by the presence of flexibility restricting parts of the bottle such as the base of the bottle. Here we consider flat glass.

Remember that glass is weak in tension and strong in compression. Consider then a pane of glass under flexion pressure held firmly at the edges (see Figure 1.5). Point A is in tension and is likely to be the origin of the breakage. The opposite side is in compression and will typically not initiate breakage (under flexion stress). Radial cracks will begin at point A and will radiate out in several directions. Points B and C are also in tension and will typically initiate concentric fractures. Hence, the "spider's web" appearance of flat glass after flexion breakage.

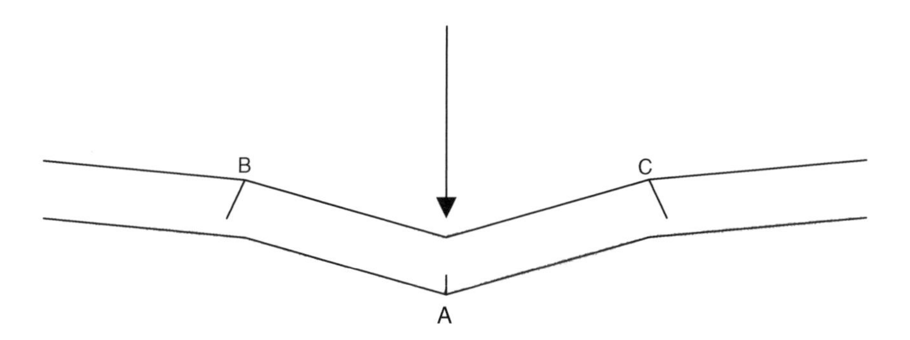

Figure 1.5 The origin of breakage under flexion.

1.7.2 Determination of side of impact

This determination is performed by an examination of ridges on the broken surfaces of the glass. These ridges are termed rib or hackle marks. According to Preston,[14] the long curved lines have been termed rib marks, while the shorter straight lines are hackle marks (see Figure 1.6).

Figure 1.6 Hackle mark comparison for individualization of fragments.

The fracture always approaches the rib mark from the concave side and leaves from the convex side. In a region of tension the fracture will travel rapidly, leaving the rib marks well spaced, whereas in an area of compression they will be closer together.

A single, well-defined rib occurs at the termination of the fracture and is a clue as to whether a fracture has extended subsequently.

To perform this examination, the glass remaining in the frame should be labeled inside and outside and much of the glass from the ground should be retained. The examiner begins by reconstructing the pane until the point of impact is obvious. The examiner should not rely on the triangular shape of a single piece or of cratering to determine the point of origin, but rather should reconstruct the pattern of radial and concentric cracks. Then the "four Rs" rule may be applied.[15,16]

Ridges on **R**adial cracks are at **R**ight angles to the **R**ear.

The origin of the rule can be seen by considering the tension and compression sides of the glass. The closely spaced rib marks are on the compression side, whereas widely spaced ones are on the tension side. Typically, the tension side is at the rear in flexion breakage for the radial cracks; therefore, the widely spaced ridges are to the rear.

This rule is unreliable on laminated glass. However, the side of impact is usually obvious in laminated glass due to the fact that it remains deformed after breakage.

Fractures in tempered glass appear to originate in the central region regardless of where the point of impact was. The surfaces of tempered glass are in compression, whereas the central region is in tension. This makes interpretation of hackle marks in tempered glass difficult.[17]

Photography of rib or hackle marks is facilitated by shadow photography.[17]

1.7.3 Percussion cone breakage

Perhaps the most obvious small, high velocity, and dense object that may strike glass is a bullet (see Figure 1.7), although cases involving sling shots and other projectiles are known.

Figure 1.7 Cratering effects on the exit side of a bullet shot through glass.

Experiments on shooting through glass show fractures both of the flexion type and the percussion cone type. "Flexure breaks, essentially radial cracks, form first and then the percussion cone is driven through to the rear side. This is attested by the fact that the conical flaw was not smooth and perfect, but made up of radial segments which terminated against the radial cracks."[12]

Care should be taken if the projectile has not penetrated the glass, as the larger side of the percussion cone may be toward the point of impact.

When two projectiles penetrate a pane of glass the sequence can be determined from the observation that the cracks for the second impact terminate at those caused by the first. Care should be taken as some cracks may "extend" by any movement after the incident.

1.7.4 Transfer of glass

In 1967 Nelson and Revell[2] reported the backward fragmentation of glass and demonstrated it by photography. Backward fragmentation is the scatter toward the direction of the force and has since become known as backscatter. They broke 19 sheets of glass and observed backward fragmentation in all cases. They observed that "whenever a window is broken, it is to be expected that numerous fragments of glass will strike any person within a few feet of the window."

The size of the fragments that are likely to be transferred from crime scenes depends on the type of glass that was broken and many other factors that are discussed later. However, there are two types of crime that have a reasonable chance of producing large fragments for comparison. These are hit-and-run incidents (incidents where a vehicle has collided with another vehicle, pedestrian, or other object) and ram raids (incidents where a vehicle has been used to effect entry). In hit-and-run cases, broken tempered glass produces relatively large pieces often scattered at (or about) the point of impact. These pieces usually include original surfaces; full thickness measurements are also possible.

Ram raids may result in large pieces of window glass being transferred to a vehicle.

However, in the bulk of cases of glass breakage, glass transfer to clothing involves small fragments with their size ranging from 0.1 to 2 mm and their shapes being typically irregular.

1.8 Physical examinations

One of the determining factors in the choice of analytical scheme employed in any given case is the size of recovered fragments. If the fragments are large, then many possibilities exist that are not realistic for the smaller fragments typically recovered from clothing. We will first present the physical examinations that can be performed on large fragments.

1.9 Examinations of large fragments

1.9.1 The comparison of thickness

Occasionally, fragments of glass are recovered that exhibit the full thickness of the source object. These can be compared by direct thickness measurement. Once measured, it may be necessary to compare two sets of measurements. Unlike refractive index, no body of statistical work exists in this area, although similarities in the comparison problem suggest that a similar solution is plausible.

It would be tempting to compare two sets of measurements directly by determining whether the ranges overlap or by use of a *t*-test or similar statistical test. However, sampling issues are quite serious with respect to thickness. Take for instance a toughened windscreen thought to have been the source of glass on a road at the point of impact with a pedestrian. The glass samples from the car are likely to be from the edges of the windscreen and those on the road from the center. Given the way in which windscreens are made and shaped, it is plausible that the edges are a different thickness, albeit slightly, to the center due to the bending in the production of the windscreen.

In this section we shall restrict ourselves to a subjective approach along the lines of "how close must two sets of thickness measurement be for them to be from the same source."

Renshaw and Clarke[18] give the standard deviation they encountered in seven vehicle windscreens as ranging from 0.004 to 0.013 mm for the five float glass samples and 0.013 and 0.037 mm for the two nonfloat samples.

This difference between float and nonfloat samples seems to also exist in flat glass samples; however, the standard deviations for flat glass are lower. We are, however, unaware of published studies on this.

Survey work[18,19] suggests that there is a very high level of discrimination between glass samples, using thickness notwithstanding, and that most glasses are sold at nominal thickness such as 3, 3.5, 4 mm, and so on. This appears to be because there is only a loose agreement between nominal thickness and actual thickness.

1.9.2 The comparison of color

Large fragments of glass can be compared for color and ultraviolet (UV) fluorescence relatively straightforwardly by subjective comparison side by side. For color, samples may be viewed in a variety of lights, and it is often valuable to use daylight and tungsten light. The background should be neutral or a complementary color (for instance, the complement of blue is orange). It is thought that a background of the complementary color "primes" the eye to observe subtle color differences.

Difficulty is found in expressing the evidential weight of these subjective comparisons, although they do have value. This is because of the difficulty in storing subjective color information from samples to produce an estimate of frequency. Attempts to overcome this difficulty, using visible spectroscopy and a microspectrophotometer (MSP), suggest limited discrimination (Newton, A.N. unpublished results, 1999).

Small fragments of glass must be treated with great caution with regard to color and UV fluorescence. It is recommended that all glass examiners should arrange for themselves to be blind tested. This can be achieved by producing casework-sized glass fragments from glass of differing colors and then mixing these fragments into debris. Most examiners will soon discover that they can barely determine a brown beer bottle fragment from clear glass or from a strong green wine bottle when the fragments are small, dirty, and among debris. Differentiating, for example, the pale greens and blues often used in vehicles is even more difficult.

1.9.3 Matching edges and matching hackle marks

With large, recovered glass samples, the fragments may first be examined to determine if coincidental edges exist between the recovered and control fragments in the form of a physical fit. This is most feasible for large pieces of glass and is seldom attempted on the typically smaller fragments transferred to clothing, although rare successes are reported. The observation of these "fracture matches" indicates that both fragments once formed a larger piece of glass.

Similarly, the edges of a broken fragment will exhibit mechanical markings, known as "hackle" marks (see Figure 1.6), which occur with such irregularity that a perfect match of an edge from a recovered fragment to an edge from a source fragment also indicates that both samples have come from the same source. The successful association by a hackle mark match is also a rare occurrence in casework and is typically not attempted on fragments recovered from clothing.

1.9.4 Density comparisons

One of the analyses that may be performed for comparison of glass samples is density. Depending on the information wanted by the forensic scientist, density can be performed in different ways: the density of the recovered and control fragment can either be compared directly or both densities can be determined. Density can be measured at a constant temperature or by varying temperature.[20] The gradient of density technique has also been suggested in literature.[21,22] As the latter technique has been shown to be more discriminating, we will present it here. The method most commonly used places a glass fragment into a liquid of approximately the same density as the glass (~2.5 g/mL). The density of the liquid is varied by the addition of a miscible liquid which is more or less dense until the fragment is observed to "float" in the liquid, not moving up or down in the tube. At that point, the density of the liquid and the glass are equal, and the density of the liquid can then be measured. The densitometer can be calibrated with glasses of known densities. The use of bromoform ($CHBr_3$) as the float liquid in conjunction with a Mettler densitometer yields a precision of +/–0.0001 g/mL.[23] An alternative to the mix of bromoform and dibromoethane as a float liquid is an aqueous solution of a nontoxic polytungstate salt. The density of this salt solution can be easily adjusted by the addition of water, and the water may be evaporated to concentrate the salt, making it reusable.

Gamble et al.,[24] Dabbs and Pearson,[25] and Miller[16] reported results for refractive index measurements and density on a set of glass fragments and noted the strong correlation.

Typical refractive index and density results for casework glass samples (see Figure 1.8) support these results. Gamble et al. note that "specific gravity was not as valuable in distinguishing glass samples as was the refractive index." However, they do go on to recommend that "it is desirable to deter-

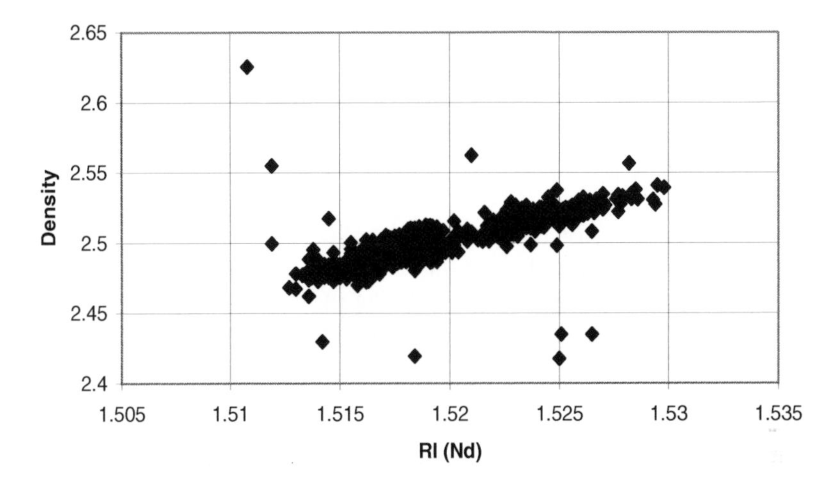

Figure 1.8 Density vs. refractive index for over 2000 case examples in the U.S. collected between 1977 and 1990. (Data from FBI laboratory, FBI Academy, Quantico, VA, Robert Koons, personal communication, 1995, by a densitometer.)

mine as many physical properties as possible." This suggests that, despite the obvious correlation, these authors felt that there was still value in determining density if refractive index determination was also performed. Stoney and Thornton[26] have shown that measuring both physical properties may or may not be useful depending on the error of the techniques.

Improvements in equipment and method, however, have increased the advantage that refractive index determination has over density, and now very few laboratories perform both. The small added discrimination is adjudged not worth the extra cost of analysis.

Interestingly, the correlation between refractive index and density can be used to refine the density method,[27] since the approximate density implied by the refractive index can be used to prepare a particularly discriminating density gradient system.

Notwithstanding these advances, the refractive index has largely displaced density as the physical property of choice to be determined. For example, a recent survey conducted by the SWGMAT* group addressed the question of added discrimination by density analysis when the refractive index determination by the Glass Refractive Index Measurement (GRIM) was performed. One laboratory that had been collecting both refractive index and density data over many years could show only one example of further discrimination by density when refractive index was also measured. This correlation also holds true for refractive index and density ranges above the usual values for casework.

* Scientific Working Group on Materials (SWGMAT) sponsored by the FBI laboratory, FBI Academy, Quantico, VA.

This finding supports the conclusion that the measurement of both refractive index and density may produce some additional discrimination, but this is limited by the observed correlation between these two properties.

1.10 Examinations performed on small and large fragments

1.10.1 Recovering glass

A number of methods for detection and collection of glass from clothing have been reported. These include hanging the garment over a large piece of paper and scraping the debris with a clean metal spatula onto the paper for collection; picking with tweezers; using adhesive tape on large surfaces (or on a surface where fibers are also gathered); in the case of the inside of a vehicle or a large room, vacuuming the area in question and examining the filter; and shaking the garment over paper or over a large (previously cleaned and blanked) metal cone. For garment examinations, each garment should preferably be packaged and inspected separately. The relative advantages of recovery methods can usefully be debated.

Pounds and Smalldon[28] give a comparison of visual searching vs. recovery by shaking. Table 1.1 gives a quick summary of these findings.

Table 1.1 A Summary of the Findings of Pounds and Smalldon[28]

Size range	Recovery by visual searching	Recovery by shaking
0.1–0.5 mm	0–12%	72%
0.5–1.0 mm	13–67%	85%
1.0 mm or greater	44–100%	93%

Interestingly, when visual searching was performed by two experienced and two inexperienced examiners, both sets performed similarly. The ranges in the visual searching column relate to the different performances of the different examiners and different garment types. These findings suggest that shaking is an efficient way of recovering glass, especially in the critical size ranges for typical casework, 0.1 to 0.5 and 0.5 to 1.0 mm. It is noted that the British view of shaking is a more violent activity than the impression given by the SWGMAT description of "scrape the debris with a clean metal spatula." Nonetheless, they both constitute shaking-type methods of collection.

A remaining point of debate is whether to shake over clean paper (and, if so, how) or a metal cone. The advantages of the cone are seen as a greater chance of recovery of fragments during violent shaking with the consequent risk of carry over and contamination. If the cone is to be used, strict anti-contamination procedures are typically required, including "blanking"* of the cone between suspects. The paper method requires that contamination does not come from the surface under the paper. This is typically achieved by cleaning this surface and, in some cases, placing a taped down clean sheet of paper on the work surface before placing the sampling sheet of paper.

* Blanking involves sampling the cone and proving the absence of carried over glass.

Fuller[29] reported recovery rates for various methods of hair combing and found variable results. Plain combs and combs treated with cotton wool and Litex were evaluated. No clear trends were discernible regarding the most efficient type of comb.

1.10.2 Examination of transparent material to determine whether it is glass

Many examiners begin their career picking up quartz and glass in equal measure from the debris produced by shaking garments. However, experience tends to allow the examiner to pick up glass exclusively by choosing those fragments with freshly broken edges and "appearance," sometimes referred to as the conchoidal fracturing.

If small particles of clear material are encountered, these can be examined under crossed polarized light conditions to determine if they are isotropic. Anisotropic materials such as quartz or other mineral crystals are birefringent and rotate plane polarized light. Therefore, they appear bright under crossed polarizers in at least some orientations. Isotropic materials such as glass would remain dark in all orientations. Plastic particles can be distinguished from glass particles by applying pressure on the particle with a hard point and determining if the material compresses.

The characteristic appearance of quartz in oil can often be recognized when the refractive index measurement is attempted, and many examiners do not take the step of determining whether or not the material is glass before proceeding to refractive index determination. It is hoped that with experience examiners do not pick up much quartz and if they do the fragments can be detected at the refractive index measurement stage.

1.10.3 The examination of surface fragments

A surprisingly high fraction of casework glass samples recovered from clothing exhibit an original surface, that is a surface that is part of the surface of the original object.[30] This suggests that examination of these surfaces might be a quick and effective way to examine glass fragments.

Elliot et al.[31] point out that examination with a standard stereomicroscope with coaxial illumination allows the examination of specular reflection, reflections from the surface of the glass. "Examination of the specular reflection image reveals surface details such as pits and scratches.... Curvature (or lack of it) is revealed by observing the reflected image of, for example, a probe tip as the probe is moved above the surface. For a flat surface, a sharp image can be obtained at every point; for a curved or irregular surface the image is distorted."[31]

This method is simple and uses available equipment; however, interference microscopy appears to offer advantages. Elliot et al.[31] describe the use of interference microscopy and their application to surface examination of glass fragments. Locke and Zoro[32] describe a method for performing this surface examination using a commercially produced two-microscope system. This methodology has not achieved widespread use despite its utility.

The phenomenon of interference involves allowing monochromatic light, which has traversed a constant distance (the reference beam), to interact with light that has reflected from the surface of a glass fragment. If we imagine the fragment to be flat and very slightly angled, then there is some line on this fragment where the light is reinforcing with the reference beam. At some distance from this line the glass is either further away or closer to the source and does not reinforce. Further away still, the distance will again become an integral number of wavelengths and will again reinforce. The effect is a series of lines of dark and light. For flat objects these lines are straight whichever way the glass fragment is oriented. For curved objects the lines are curved in most orientations.

Locke and Elliot[33] pursue a classification approach with much success. Using their recommendations, glass is classified as flat, curved, or slightly curved or by suggesting sources such as inner surface of container or tableware,[a] outer surface of container or tableware,[b] spectacle lens,[c] nonpatterned surface of patterned glass,[d] patterned surface of patterned glass,[e] flat glass,[f] or windscreen glass.[g] It can be seen that some of the characteristics described would apply to two categories; however, blind trials suggest a very high success rate once experience has been obtained. Error rates are higher for small fragments. This suggests that classification should be applied to fragments 0.4 mm^2, or to sets of three fragments. If sets of three fragments are to be used, then they should all be <0.4 mm^2 in size and they should be similar in refractive index and interference pattern.

Experience suggests that the technique has discrimination beyond that suggested previously. The flat sides of different one-sided patterned windows are easily differentiated, for instance, due presumably to the different surface defects on the flat roller. Even flat glass shows differing amounts of surface markings from cleaning and may be discriminated in some cases.

Given the cheapness and effectiveness of this technique, it is surprising that more use is not made of it. It also seems likely that a high fraction of surface fragments is an indicator of backscattered glass. There is obvious, but unexplored, potential in this, using the Bayesian approach, to help distinguish between some prosecution and defense hypotheses.

As previously suggested, float glass fluoresces when excited at 254 nm.[31] This fluorescence gives the surface a dusty or milky appearance that may appear faintly yellow or green. This fluorescence is not necessarily visible

[a] Smooth curved lines due to the blown nature of the inner surface.
[b] Wavy curved lines due to the mold contact of the outer surface.
[c] Curved straight lines.
[d] Straight, but wavy lines resulting from flat surfaces affected by contact with the roller.
[e] Complex wavy lines.
[f] Smooth straight lines in all orientations.
[g] Smooth, very slightly curved lines.

in casework-sized fragments due to their small size and reflections from broken surfaces.

Lloyd[34] reports on experiments examining the fluorescence spectrometry of glass surfaces of casework-sized pieces. The main emission from a float surface is a very broad band with a maximum at about 490 nm when excited at 280 nm. Another excitation occurs at about 260 nm and produces a second broad emission largely superimposed on the first. A third excitation occurs at 340 nm, producing an emission peaking at 375 nm.

Nonfloat surfaces show only one strong fluorescence with an excitation at about 340 nm and an emission peaking at about 375 nm, similar to the third excitation of the float surface. The excitation at 280 nm is thought to result from tin. Considerable variation occurs between the fluorescence intensities for different samples of float glass. The largest variation lies in the tin fluorescence.

Locke[35] describes equipment for viewing the float glass fluorescence on casework-sized fragments. Inquiries suggest that this equipment is not commercially available. The float surface may also be detected by the use of X-ray analysis using the SEM to detect tin.

1.10.4 Refractive index determinations

The dominant physical property investigated in glass has been refractive index (RI) measurements. RI can provide a high degree of discrimination between a known and a questioned glass sample.

RI measurements are widely used for comparing forensic glass samples in crime laboratories. This optical property is easily measured in transparent materials such as glass and has been measured for more than 60 years with various methods. As such, it has largely superseded density as the physical property of choice to be measured.

Figure 1.9 illustrates the change of direction (refraction) that is observed when light passes from one medium to another. Refraction occurs because (as described by Snell's Law in Equation 1.1; also see Appendix A) the velocity of light in the transparent medium is slowed.

This interaction can be described as (1) the ratio of the wave's velocity in a vacuum to the wave's velocity in the transparent medium or (2) the ratio of the sine of the incident angle to the sine of the angle of refraction.

$$\mathrm{RI} = \frac{\sin\theta_I}{\sin\theta_R} = \frac{V_{Vacuum}}{V_{Glass}} \tag{1.1}$$

Becke* first described a phenomenon that he observed while analyzing geological samples in 1892. Using a microscope to examine rock sections, Becke observed a bright line inside the edge of a mineral section that had a

* Friedrich Johann Karl Becke, mineralogist, 1855–1931.

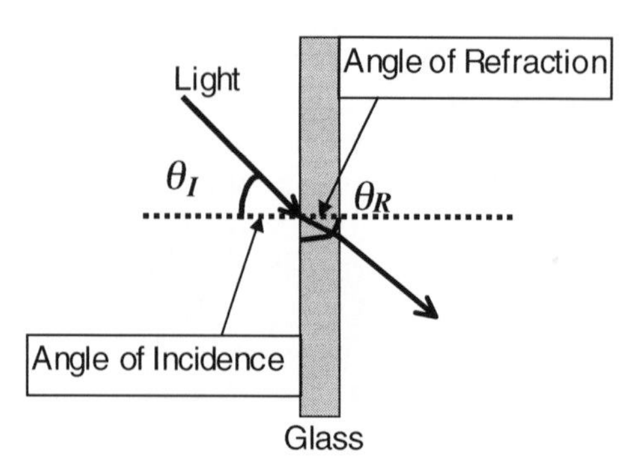

Figure 1.9 Refraction of light through a glass medium.

higher RI than its surroundings. Becke also saw that when the objective on the microscope was raised (focus up), the bright line moved in the direction of the mineral of higher RI. This Becke line, as it is known today, is a function of how light behaves at the boundary between two components with different RIs. To measure the RI of glass, one could immerse the glass in oil and adjust the focus on the edge of the glass fragment. If the objective was raised and the bright line moved into the oil (and away from the edge of the glass), then this would indicate that the oil had a higher RI than the glass. One could then take a miscible liquid such as an organic solvent and add it to the oil to lower the RI line until it does not move further or disappears. At that point, the RI of the oil is equal to the RI of the glass. One could then measure the RI of the oil with a refractometer. This method is very time consuming and not very accurate.

Around 1930, Emmons described the temperature variation technique, observing that RI varied with temperature. As the temperature of an oil is raised, the RI of the oil decreases while not affecting the RI of the immersed glass to the same extent. Emmons used a hot circulating water bath to heat the microscope slide and its contents and then waited for the Becke line to disappear. The water bath was also used to heat a sample of the same oil on a refractometer so that the RI could be measured once the line disappeared. This method was an improvement over the previous Becke method, but it was hard to control the temperature accurately. Therefore, the error was relatively high.

The double variation method[16] involved the variation of both the temperature and wavelength of the light coming through the sample to determine the RIs for three wavelengths. This method has been published as an accepted method by the Association of Official Analytical Chemists.

Mettler developed and sold a commercial Hot-Stage™ in the early 1970s that enabled very good control of the temperature of the locally heated microscope slide and was typically used in conjunction with a phase contract objective. This method was more accurate than previous methods, but still very time consuming. The phase contrast objective[36] enhances any phase contrast caused by retardation or acceleration of the light as it passes through the glass relative to light that has passed only through oil. This is achieved by adding or subtracting a large phase difference (typically $\lambda/4$) to the small phase difference existing between those rays which have interacted with the sample and those which have traversed the surrounding border. This enhanced phase difference is seen as an optical effect similar to the Becke line, that is light and dark lines, but is strictly not the Becke method since that method uses a defocusing which is not required in phase contract microscopy. The act of defocusing and refocusing becomes progressively harder as the match temperature is approached and the fragment becomes less visible.

As noted by Ojena and De Forest,[3] "it becomes increasingly difficult to see the Becke line as one gets closer to the match point of the immersion liquid and the glass fragment. The contrast of the diffraction pattern (Becke line) is influenced by the shape and dimensions of the specimen and by the degree of defocusing of the microscope. These factors, in turn, affect the accuracy of the match."

In the mid 1980s, Foster and Freeman developed an instrument which they called "Glass Refractive Index Measurement" (GRIM) based on the observation of the glass immersed in oil (Figure 1.10).

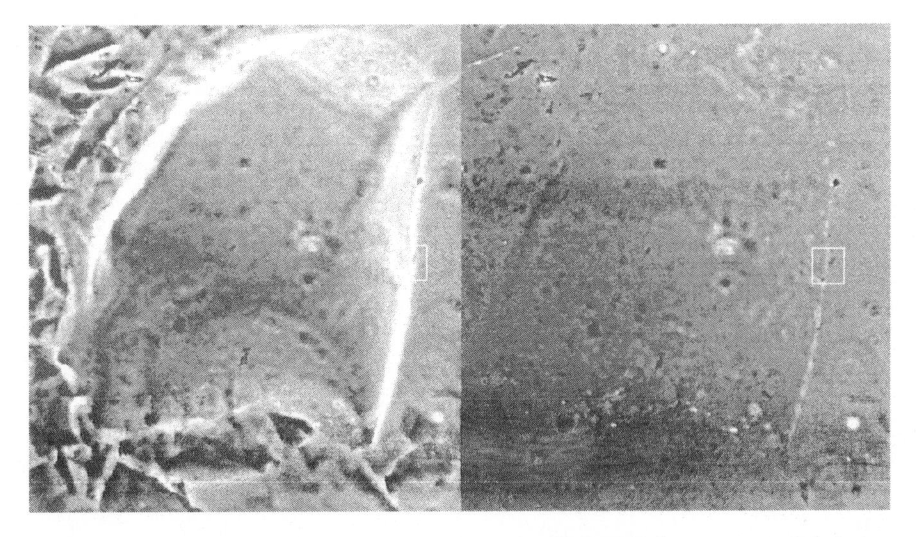

Figure 1.10 Glass fragments as viewed from the GRIM II video camera while being immersed in the Locke B Oil. The fragments on the right are invisible due to the fact that they are at the "match temperature."

A phase contrast microscope is used at the fixed wavelength of 589 nm, and the image is adjusted for maximum contrast. GRIM I and II both use dark contract phase microscopy. This is where the diffracted rays, those that have interacted with the sample, are deliberately retarded with respect to the 0-order (undiffracted) rays.

Three studies have been conducted to evaluate the accuracy, precision, and long-term stability of the GRIM method and conclude that the instrument provides for very satisfactory results and should be the method of choice for the measurement of RI in forensic laboratories.

The GRIM manual suggests that fragments producing edge counts below ten should not be used or should be repeated manually. Tests by Zoro et al.[37] and Coulson, Curran, and Gummer (personal communication) suggest that there is an increase in the variation of measurement on a single glass pane with a lower edge count, but the mean remains approximately the same. This suggests that low edge count fragments may still be used if the added variance is considered in the comparison and assessment procedure.

1.10.5 Dispersion

RI varies with wavelength, being greater for shorter wavelengths.[16] The standard wavelength for RI determination is the sodium D line. If no superscripts or subscripts are used, then the RI referred to is the Sodium D line at 25°C. Therefore, N is shorthand for N_D^{25}. Other standard wavelengths used are near the ends of the visible spectrum (typically the hydrogen C [656.3 nm] and F [486.1 nm] lines). The RI at these wavelengths are N_C and N_F, respectively.

There are two ways of expressing dispersion. First, it may be shown as a plot of RI vs. wavelength or, second, it can be defined as

$$V = \frac{N_D - 1}{N_F - N_C} \tag{1.2}$$

Locke et al.[38] suggest that "for glasses of the type commonly encountered in casework in the United Kingdom (of Great Britain and Northern Ireland), dispersion measurements are unlikely to enhance the evidential value" due mostly to the large observed correlation between dispersion and N_D. Locke et al.[38] and Cassista et al.[39] dispute the findings of Miller[40] and Koons et al.[41] who suggested the opposite conclusion. However, in Miller data have not been given on the discrimination added, and in Koons et al. no phase contrast has been used, suggesting the possibility of low precision.

1.10.6 Refractive index anomalies

It has been known for some time that the float surface of float glass has a different RI to the bulk of the glass. Underhill[42] describes the origin of the multiple RI phenomena. The float surface is enriched in SnO_2, and the opposite surface appears to be depleted in various other constituents and, consequently, enriched in SiO_2. The float surface typically has a lower RI to the bulk. Underhill[42] and Davies et al.[43] raised the exciting prospect that a separate float RI might be measurable and, hence, an additional aspect of discrimination might be available. Later research[37] confirms the presence of RI anomalies at the float surface, but suggests that this RI is approached smoothly over the last few microns of depth and, therefore, a discrete surface RI does not exist. They also show that the RI of the antifloat surface (the surface opposite the float surface) has an anomaly in the other direction (that is toward higher RI). This antifloat anomaly is smaller than the float anomaly in most, but not all, of Zoro et al.'s samples.[37]

Nonfloat flat glass may also on occasion exhibit surface anomalies. Two of Zoro et al.'s nonfloat samples showed anomalies, in both cases toward higher RI. One sample showed this at one surface and another at both surfaces.

Container glass and tableware may also show surface anomalies (in this case in either direction).

These findings do not absolutely exclude the use of "surface RI" as a discrimination tool, but do suggest that this use will be very difficult. It is, for instance, theoretically possible to consider "maximum" deviation as a discrimination tool. In order to do this it will be necessary to identify fragments expected to show maximal deviations, possibly by confirming that the measured edge is truly surface (and reasonably thin but measurable). The authors are aware of no laboratories applying this technique.

1.10.7 The examination of tempered (toughened) glass by annealing

Locke et al.[44] describe an annealing process where a specifically designed oven is used to anneal the small fragments typically recovered from clothing. Locke and Hayes[45] report the results of this technique when applied to a number of different glass sources.

Table 1.2 is a summary of Locke and Hayes findings; positive ΔRI values mean that the RI was higher after annealing.

Table 1.2 A Summary of Locke and Hayes'[45] Findings

Glass type	ΔRI
Tempered specimens	0.00173–0.00206
Nontempered float, patterned, or plate glass	0.00086–0.00144
Container glass	0.00073

Table 1.2 suggests that tempered specimens can be differentiated from nontempered specimens. Later work[46] on 200 random survey items suggested that an ΔRI value above 0.00120 identifies the glass as toughened, while values less than 0.00060 identify the glass as not toughened with an inconclusive region in between (please note that these values do not follow from Locke and Hayes' work which would suggest a value of ΔRI = 0.00150).

The studies performed by Cassista and Sandercock,[39] Locke et al.,[44,45,47,48] Winstanley and Rydeard,[49] and Marcouiller[50] confirm that toughened glass can successfully be classified using annealing.

This approach is the traditional classification approach. A different approach is possible. The ΔRI value per se is an element of discrimination and can be used as that rather than as an element of classification. This point is a theme in glass where much effort has been expended in classification type investigations where the discrimination type approach is as valid if not more so. In the case of annealing, two types of discrimination information are available. The first described earlier is the ΔRI value, and the second arises from the fact that the variance of RI in annealed glass is less than in unannealed glass. Hence, the comparison of two sets of appealed measurements is expected to give more discrimination (or a more peaked density, see later).

1.11 Elemental composition

Elemental analysis can either be used for classification or to discriminate glasses of the same type (windows, container, etc.) having similar physical characteristics. Indeed, RI differences correspond to small compositional differences in the major elements silicon (Si, ~30%), sodium (Na, ~8%), calcium (Ca, ~8%), magnesium (Mg, ~2%), and potassium (K, ~1.5%). It is possible, however, to have two different glass samples and be able to observe the same density and RI, and measure a difference in the minor element composition of aluminum (Al, ~1%) and iron (Fe, <0.3%), or by trace element content of barium (Ba), manganese (Mn), titanium (Ti), strontium (Sr), and zirconium (Zr). Ba, Mn, Ti, Sr, and Zr are all usually present in concentrations of less than 0.1%.

The techniques used for classification and discrimination may be different, as the determination of glass type depends on the major elements, whereas discrimination is mostly based on minor or trace elements. Numerous methods have been proposed for the elemental analysis of glass, such as SEM-EDX,[51-53] X-ray fluorescence (XRF),[54-57] neutron activation,[58] or inductively coupled plasma (ICP) techniques.[41,59,60] Many different techniques for the analysis and comparison of glass samples have been used because in many cases forensic laboratories must often use the available instrumentation and adapt its use to as many types of evidence analyses as possible. A review of some of the methods used for the elemental analysis of forensic glass samples was reported by Buscaglia[61] who discussed the advantages and disadvantages of the individual analytical techniques.

Although each of these techniques has its strengths, each is limited in practical usage by one or more of the following: sample size requirements, whether or not the technique is destructive, sensitivity, precision, multi-element capability, or analysis time.

It has been argued that the last four decades have seen a change in the glass manufacturing industry as it greatly improved the quality control within glass plants due to the introduction of the float process for glass manufacturing. This improvement in the quality of the glass product has been further enhanced by computer-managed processes that also allow manufacturers to control the physical and optical properties of the glass such as thickness and RI to a great degree. In addition, the methods and formulation among manufacturers and plants around the world has become more uniform. This has caused a noticeable decrease in the variation of RI for the population of flat glass produced by any one plant. While this trend appears to have been at least partially offset by the growing internationalization of the market, leading to more plants selling into any one market, there have been persistent concerns about a general loss of overall discrimination if the sole method used is RI. This has led to a critical review of other methods of analysis that could add discrimination.

A second school of thought considers elemental analysis as a tool to be used when required. Therefore, depending on the circumstances, there may be no need for elemental analysis or there may be a need for a technique that adds a little more discrimination or a very powerful technique. The scientist will not only consider the discriminating power of the technique when choosing if and how elemental analysis should be performed, but she/he will consider the circumstances of the case, the cost, the destructive nature of some techniques, and the availability of equipment. It is in this context that we would like to present the techniques used nowadays in forensic laboratories or in research: X-ray and ICP techniques.

1.11.1 X-ray methods

Two techniques have been proposed for the analysis of glass: SEM-EDX and XRF. Both are considered to be rapid, relatively sensitive, nondestructive, and complementary, as SEM is more sensitive to small atomic number elements (major and minor elements in glass). These techniques have both been used for classification and discrimination.

1.11.1.1 Classification of glass using X-ray methods

Keeley and Christofides[56] showed that magnesium and iron allowed the differentiation of windows from other glass types. These results have been confirmed by other researches using SEM-EDX, such as Ryland.[53] Ryland used both SEM and XRF, as the latter is more sensitive to iron. In this research, it was found that old windows contained low levels of magnesium, but that the high iron content still allowed differentiation from container glass. Terry et al.[52] used sodium, aluminum, silicon, calcium, and potassium in addition to magnesium. The content of potash and lime allowed the differentiation of borosilicate labware from borosilicate headlamps; the content of potash (in excess of 5%) differentiated spectacles and lead glasses. Flat glass was identified by its low aluminum content. Howden et al.[54] successfully used XRF to classify glass; the elements used were sodium, magnesium, potassium, calcium, iron, arsenic, barium-titanium, strontium, and zirconium.

For additional information on classification schemes showing differences between container and float glass, the reader is referred to studies conducted by Hickman.[59,62,63]

1.11.1.2 Discrimination of glass using X-ray methods

XRF techniques have also been reported to discriminate glass of the same type, having indistinguishable RI and density. In the first study published on the subject, Reeve et al.[51] reported that out of 81 glass samples, 2 were indistinguishable. In a similar study, Andrasko and Maehly[64] were able to distinguish all but 2 of 40. With emission spectroscopy a further pair was discriminated. Studies on the discriminating power of the X-ray methods show that XRF may be more discriminating than SEM, but less discriminating than ICP techniques.

Rindby and Nilsson[57] have reported the use of micro-XRF. This technique as well as total reflection XRF (TXRF) seems promising.

1.11.2 ICP techniques

Inductively coupled plasma-atomic emission spectrometry (ICP-AES) has been applied extensively to trace element analysis of glass.[41,59,62,65-69] ICP-AES combines multiple element detection capability with good sensitivity, small sample size requirements, and a large linear dynamic range, making it well suited for the trace element analysis of glass.

1.11.2.1 Classification of glass using ICP techniques

ICP-AES has been proposed as a classification and discrimination method. Hickman[59] reported a scheme for classifying glass samples as sheet, container, tableware, or headlamp. RI, as well as elemental concentration (manganese, iron, magnesium, aluminum, and barium), allowed the classification of glass samples 91% of the time. In 1988, Koons et al.[66] were able to classify 182 sheet and container glasses correctly, with the exception of 2 windows classified as container and 2 containers classified as sheets. In most instances it was even possible to correctly associate each sample with a manufacturing plant.

Almirall et al.[60] have also shown the use of classification methods for glass sources with discriminant analysis routines using elemental data by ICP-AES.

1.11.2.2 Discrimination of glass using ICP techniques

Early work by Hickman et al. in the U.K. provided a basic set of elements which allowed a high degree of discrimination between glasses.[59,62,70] These elements included lithium, aluminum, magnesium, manganese, iron, cobalt, arsenic, strontium, and barium. Investigations by Koons and co-workers in the U.S. have compared ICP-AES to RI and energy dispersive XRF and demonstrated on a sampling of 81 tempered sheet glasses a much higher degree of differentiation for ICP-AES.[41] The elements measured were aluminum, manganese, iron, magnesium, aluminum, barium, calcium, strontium, and titanium. Both ICP-AES and RI were also studied extensively by Almirall in his Ph.D. thesis. Wolnik et al.[67] reported that the use of ICP-AES for the discrimination of container glass using barium and magnesium concentration proved to be useful to discriminate manufacturers.

A more sensitive and potentially more informative method of glass analysis is inductively coupled plasma-mass spectrometry (ICP-MS).[71,72] Analytical applications of ICP-AES and ICP-MS are similar, while ICP-MS usually affords 10 to 100 times the sensitivity of ICP-AES, allowing either ultratrace element detection or the analysis of smaller samples. ICP-MS also provides isotopic information that is useful in isotope dilution analyses, which improves both accuracy and precision.[73]

The first reported application of ICP-MS to forensic glass samples was made by Zurhaar and Mullings[73] who analyzed seven glasses with identical RIs. Due to the sensitivity in ICP-MS, samples as small as 500 μg were analyzed with detection limits below 0.1 ng/mL. While 48 elements could be accurately determined with relative standard deviations of less than 4%, it was found that a suite of 25 elements produced successful discrimination levels in 85 to 90% of the glasses tested.

A more extensive investigation of ICP-MS analysis of glass has recently been published by Parouchais et al.[74] Sample digestion methods were compared, and up to 62 elements were determined in a range of glass samples. Successful differentiation of glasses of similar RI was accomplished by comparing element concentrations and element ratios (e.g., Sr/Ba). The latter technique is useful in that samples need not be weighed, thereby simplifying the analysis procedure. The use of elemental ratios is appealing because they are not dependent on mass, not subject to instrument drift, and are less prone to signal suppression effects. However, they can be subject to temporal variations due to changes in mass bias and plasma temperature, making database development difficult.

An Australian group (personal communication, the National Institute for Forensic Science, NIFS) and the BKA (BundesKriminalAmt) have conducted research on the application of laser ablation (LA)-ICP-MS. This method has the great advantage of being fast and nondestructive.

These studies have demonstrated that trace element analysis by ICP-AES and ICP-MS is a powerful method for discrimination of forensic glasses when more common techniques, such as RI measurement, fail to discriminate.

Interpretation of elemental concentrations, whether determined by XRF, ICP-MS, or LA-ICP-MS, requires a fundamental knowledge of the variation within a population. The fundamental knowledge of these distributions is not presently available for ICP-MS. It is expected that current research in the U.S. and Germany will take the elemental analysis of glass by ICP methods from the research laboratory to the operational crime laboratory and hence into the courtroom.

1.12 Summary

We have presented a sampling of techniques available to characterize glass. The main thesis of this book is how to interpret the results produced. From here onward, we will be presenting both the "classical approach" to interpreting this data and the "Bayesian approach," which we advocate. It was thought necessary here to introduce the type of data that may be developed on glass so that some, at least, of the subtleties might be understood.

1.13 Appendix A — Snell's law

The relationship described by Snell's law is easily derived. Construct the triangles shown by dropping perpendiculars. Note that the angles marked are θ_i and θ_r by using the concept that internal angles of the triangles are θ_i, 90, and 90 – θ_i and θ_r, 90, and 90 – θ_r, respectively. Then,

$$\sin\theta_i = \frac{x}{h} \tag{1.3}$$

$$\sin\theta_r = \frac{y}{h} \tag{1.4}$$

and, therefore,

$$\frac{\sin\theta_i}{\sin\theta_r} = \frac{x}{y} \tag{1.5}$$

and x/y are in the ratio velocity in a vacuum/velocity in the medium.

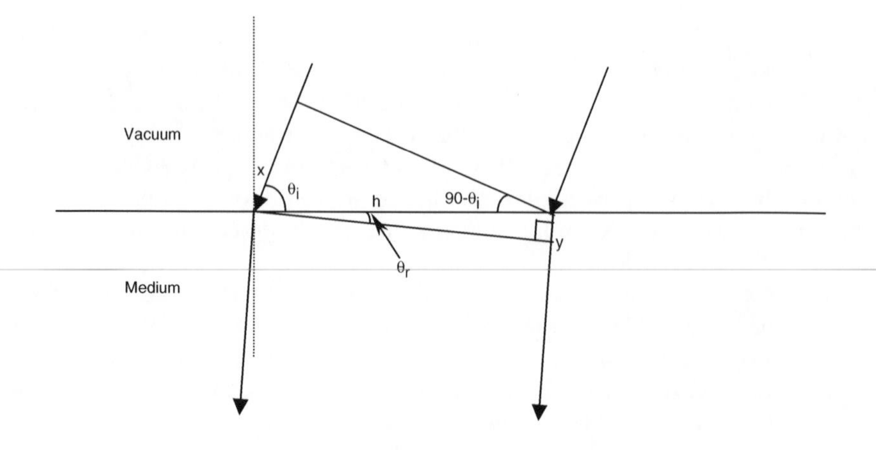

chapter two

The conventional approach to evidence interpretation

Let us assume that the police have taken a sample of size n control fragments of glass from the crime scene and that the laboratory analyst has recovered m fragments from the suspect's clothing (and hair). A single RI measurement has been made for each fragment using the standard immersion test.[3] Therefore, we have n measurements of RI on the control glass $x_1, x_2, \ldots, x_n$ and measurements $y_1, y_2, \ldots, y_m$ on m fragments recovered from the clothing (and hair). Because of internal variation and the error of the technique, even if the control and recovered glass have the same origin, there will be differences between the two sets of measurements. Therefore, it is necessary to determine if these differences are due to internal variation and the error of the measurement technique or to the fact that the fragments come from different sources.

How should the problem of comparing these two sets of measurements be approached? At one extreme there is the view that an experienced forensic scientist can effectively do this purely by personal assessment. The other extreme calls for an objective statistical test giving a match/no match result quite independent of human judgment. Inevitably, neither extreme is sustainable in practice. The "personal judgment" school is open to the criticism of being unscientific unless the persons who make the judgments can show a convincing level of performance in tests designed specifically for the purpose of establishing competence in this activity. The "objective test" extreme follows a common illusion, regrettably fostered by many statisticians, which ignores the fact that no test can be formulated without an underlying theoretical model. The validity of such a model in any given case must be determined by personal assessment, taking account of information other than that contained in the measurements alone.

The optimal path must fall somewhere in between these two extremes. The central role of expert judgment is not denied, but it is recognized that it cannot be unfettered. It should be augmented and circumscribed by systems for ensuring competence. Neither can the enormous advances that have been made by statisticians in developing methods for use in circumstances such as these be ignored.

The next section follows the development of statistical tools in glass evidence interpretation from a historical perspective. This chapter will start from the objective test extreme, but will later develop to show how a more optimal approach may be achieved.

2.1 Data comparison

The comparison of RI/elemental measurements has been done in various ways. There are very simple tests such as the "range test" or the "±3SE rule" and some more complex tests such as the Student's *t*-test, the Welch test, or Hotelling's T^2. These methods all test the hypothesis H_0 that the recovered and control fragments have the "same" physical or chemical characteristics. However, we know that there will often be cases in which the recovered fragments have come from two or more sources. Therefore, before testing H_0, we could adopt one of two policies. The first, and oldest, approach is to test each recovered fragment against the control mean. This approach has its merits, but these are outweighed by two problems: (1) each single test is comparatively weak, and (2) the potential for drawing an incorrect conclusion increases with every test. A more powerful approach is to adopt some kind of preliminary procedure for identifying groups among the recovered fragments. The conventional methods — illustrated for RI measurements — can be schematized as shown in Figure 2.1.

It is perhaps surprising that there are so many different methods for the interpretation of glass evidence. Therefore, it is logical for one to wonder which test should be chosen and whether or not the recovered fragments should be grouped. To aid this process we will present the advantages and disadvantages of the methods.

2.1.1 Range tests and use of confidence intervals

The range test is a nonparametric test (the expression "nonparametric" means that the test design makes no assumptions about the underlying distribution of the data). It has the great advantage of being a computationally simple test. One may accept (or fail to accept) the hypothesis that the recovered and control fragments have the same physical or chemical characteristics if the measurement made on each recovered fragment lies within the range defined by the maximum and minimum values observed in the control sample. The mathematical relationship can be formalized as follows:

$$\text{Reject } H_0 \text{ if } \min_{i=1\ldots n} x_i > \min_{j=1\ldots n} y_j \text{ or } \max_{i=1\ldots n} x_i < \max_{j=1\ldots m} y_j$$

$$\text{Otherwise fail to reject } H_0$$

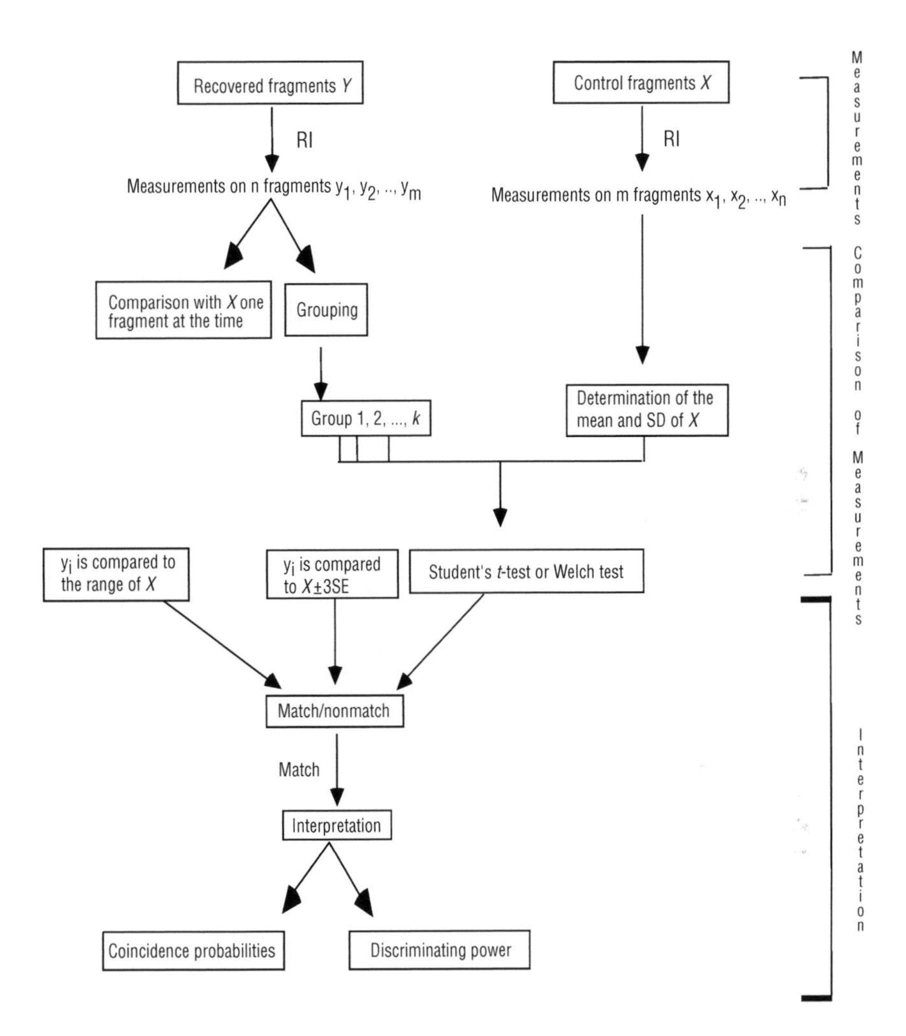

Figure 2.1 Conventional approach to glass evidence.

The real disadvantage of this method is that it has an unacceptably high false acceptance rate, i.e., it will imply two samples are the same when they are often not.

2.1.2 *Confidence interval*

The use of the confidence interval (also called of the ±2 or ±3SE rule) presupposes that the measurements are normally distributed and that the variance of the recovered glass is the same as the control glass. The test is as follows: if the measurement of the recovered fragment falls within the confidence limits determined by the control sample, then the hypothesis H_0 that the recovered and control fragments have the same physical or chemical characteristics is accepted. Otherwise, the fragment is said to be different. The mathematical relationship can be formalized as follows:

$$H_0 \text{ accepted if}: \; y_j \in \left[\bar{x} - t_{n-1}(\alpha/2)se(\bar{x}) \forall j = 1, \ldots m\right]$$

$$\text{Otherwise reject } H_0$$

where $t_{n-1}(\alpha/2)$ is the $100(1-\alpha/2)\%$ critical value for the t-distribution on $n-1$ degrees of freedom, and $se(\bar{x})$ is the standard error of the mean.

The significance level* (α) can be set at any level, but is usually one of 5, 1, or 0.1%.** There is a considerable amount of arbitrariness about the choice of significance level. Why should we choose any particular significance level — 5, 1, or 0.1%, or any other? Let us leave this issue to one side for the time being and proceed with a 1% significance level. By setting 1% as the desired significance level we accept that, in theory, on average 1% of cases in which the control and recovered fragments truly came from the same source we would incorrectly reject the null hypothesis. That is we would mistakenly conclude that the recovered fragments had not come from the window at the scene. The significance level is also known as the Type I error rate, the theoretical probability of a false negative. If we were dissatisfied with this false negative error rate, then we could change to a 0.1% significance level. However, the consequence would be that we would be less likely to discriminate in those cases in which the control and recovered fragments had, in fact, come from different sources. That is we would increase the Type II error rate or the false positive rate. Thus, the significance level is inevitably a compromise between Type I and Type II errors.[75]

Two rules that seem to avoid setting the significance level are the ±3SE rule (see, for example, Slater and Fong[20]) and the ±2SE rule.[62] However, these rules still use a significance level. The problem is that it changes with respect to sample size. For example, with five control fragments the Type I error rate would be about 4% for the ±3SE rule and 11.6% for the ±2SE rule. With 20 control fragments these figures would be 0.7 and 6%, respectively. Compared to the range test, using the rule of ±3SE presents about two times less false negatives, but three times more false positives.

* Most statistics tests such as Student's t-test or the Welch test presented later presuppose that the significance level is fixed.

** The significance levels referred to here are "theoretical." They will be correct if a number of assumptions are satisfied. In practical glass work there may be a considerable difference between the theoretical significance level and that achieved in practice. The consequence of this is that the Type I error rate (false negatives) is typically larger than the theoretical Type I error. Some effort has been made to identify which of the typically made assumptions are being violated in practical glass work, and, while there are some candidates, there is not yet international agreement on this matter.

It is very important to note that these error rates are *per test*. Each time a fragment is compared to the confidence limits of the control sample there is a probability α of drawing an incorrect conclusion. Therefore, the probability of making at least one incorrect conclusion if we are comparing m fragments is approximately $m\alpha$. For example, with a significance level of 1% and ten recovered fragments the overall Type I error rate is approximately 10%, i.e., ten times higher than desired.

2.2 Statistical tests and grouping

As the tests presented in this section assume that groups and not single measurements are compared, the different articles on grouping will be presented before we proceed with the statistical tests.

2.2.1 Grouping

There are various ways in which grouping can be carried out.[76] These can be divided into the agglomerative method of Evett and Lambert[75] and the divisive method of Triggs et al.[77,78]

Any thorough statistical analysis must allow for the possibility that recovered fragments may have come from different sources. Current treatments[7,8] of forensic glass evidence require prior knowledge of the number of groups of glass on the clothing.

Conventional statistical clustering or grouping techniques rely on large sample sizes to achieve their results. In a forensic examination, although the number of recovered fragments may be high, it is very uncommon for more than 20 of these to be measured. Thus, conventional techniques are not feasible. It is desirable to have the optimal approach for identifying groups within the recovered sample. This section will explain the divisive method of Triggs et al.[77,78] for detecting groups and demonstrate its efficacy against the currently used agglomerative methods.

In the following section we assume that a sample of m fragments of glass has been recovered and their RIs $y_1, y_2, \ldots, y_m$ have been determined. Furthermore, we assume that each fragment is drawn from one or more distinct sources. However, the true source of each fragment is unknown. The grouping problem is equivalent to assigning each fragment to a group so that those within a group have similar RIs.

2.2.1.1 Agglomerative methods

The agglomerative solution to this problem is a "bottom-up" approach. Initially, the fragments are considered to have come from m different sources or belong to m distinct groups. The agglomerative algorithms attempt to classify the fragments as having come from fewer than m sources by considering the effect of a fragment on the range of a particular grouping of fragments. If this range is too large, the fragment is put into a separate group.[75]

Evett and Lambert 1 (EL1)

The *m* RIs are ordered, and the median RI is identified along with its nearest neighbor. If these two fragments are close enough together, i.e., if their range (the largest RI minus the smallest RI) is less than some critical range $r(\alpha;2)$, then they are assumed to have come from the same source. The next closest neighbor is selected, and again the question of whether it is close enough, in terms of range of the three fragments, is asked. If the fragment is close enough, then it too is classified as having come from the same source. This process continues until either all the fragments belong to the same group or a fragment is too far away. If a fragment is too far away, then the process is repeated on each remaining set of ungrouped fragments. If there is only one fragment under consideration, i.e., all other fragments have been grouped, this fragment is said to have come from a different source and the process stops. EL1 is covered in more detail in Appendix A.

The critical values, $r(\alpha, m)$, are scaled standardized normal ranges. The scaling factor used by Evett[79] is a weighted estimate of the unknown standard deviation obtained from 230 control samples. The critical values are listed in Table 2.5 in Appendix B.

In subsequent analysis, Evett and Lambert[75] assume that the recovered samples come from normal distributions. It seems sensible then to use percentage points from the empirical sampling distributions of the range of a sample of size *m* from a normal distribution rather than the ranges used by Evett.[79] These values are listed in Table 2.6 in Appendix B. This modified algorithm is denoted Evett Lambert Modification 1 (ELM1) and has one additional change in that it checks the range of the whole sample against the particular critical value before stepping into the regular EL1 scheme. ELM1 is described in more detail in Appendix A.

A further modification to ELM1 is proposed with an alternative starting procedure to avoid the possibility that the algorithm may start on the edge of a group. This modified algorithm is denoted ELM2 and is detailed in Appendix A.

2.2.1.2 *Divisive methods*

The divisive solution to the grouping problem is a "top-down" approach. Initially, the fragments are considered to have come from one source or all belong to one group. The divisive algorithms consider the differences between the fragments' RI and whether partitioning the fragments into two separate groups would increase the homogeneity within groups.

The modifications of Triggs et al.[76-78] to the divisive method of Scott and Knott[80] gave rise to the following algorithm.

Scott Knott Modification 2 (SKM2)

The *m* RIs are ordered, and m–1 different groupings are considered. The distinctiveness of a particular grouping is assessed by comparing the summed squared distances of each fragment from the mean of the group to which it is assigned to the squared distance between the group means. Initially, we assume that the RIs of the recovered fragments are normally distributed with common variance s^2 and that we have an independent estimate, s_0^2, of s^2. The maximum summed squared distance is scaled by s_0^2 to give λ_* and compared to a critical value, $\lambda_*(\alpha;m)$ If $\lambda_* > \lambda_*(\alpha;m)$, then the fragments are divided into two separate groups, and the group membership is provided by the particular configuration that gave λ_*. If a division was made, then the process is repeated for each of the two subgroups until no further divisions can be made. The algorithm SKM2 is covered in more detail in Appendix A.

The critical values $\lambda_*(\alpha;m)$ are found by simulation. Assuming that all m fragments come from one source, 10,000 random samples of size m are drawn from a normal distribution with mean 0 and s^2. For each sample, the value of λ_* is calculated. The 10,000 λ_* values were sorted and $\lambda_*(\alpha;m)$ is the 10,000 $\times$ α largest value, e.g., for $m = 10$ and $\alpha = 0.05$ the 10,000 $\times$ 0.05 = 500th largest value of λ_*, which is 17.54. This is taken as the critical value. We expect that when the ten fragments all come from a single source, the resulting λ_* will exceed $\lambda_*(0.05;10) = 17.54$ in only 5% of cases.

The critical values $\lambda_*(\alpha;m)$ for a given m and α are listed in Table 2.4 in Appendix B. These methods can be best demonstrated by the following example.

Example

Assume that a sample of 11 fragments is recovered from the suspect's clothing and the RIs have been determined. The fragments are sorted into ascending order and have been labeled 1 to 11. It is assumed that the recovered fragments come from three different sources.

The divisive algorithm proposed by Triggs et al.[76-78] consists of a sequence of partitions of the data into two groups based on a maximal between group sum of squares (BGSS). In Figure 2.2 the divisive procedure considers the m–1=10 possible partitions of the ordered data into two groups. The BGSS is calculated for each possible partition, and the maximum BGSS is found. This maximum occurs when the data is partitioned into a group containing fragments 1 to 8 and a second group containing fragments 9 to 11. This procedure is repeated for each subgroup until no further partitions can be made. The groups are taken to be the terminal nodes or leaves of the tree in Figure 2.2. In some sense the divisive algorithm can be considered a top-down approach.

The agglomerative algorithm on the other hand works from the bottom up. The median of the fragments is determined, and the fragment with the closest RI to the median is selected as the first element of the "group." Neighboring fragments are then added into the group until their range exceeds tabulated critical values for the range of a normal sample. In Figure 2.3, the median is fragment 6. The next fragment closest to 6 is 7. This is added to the group as the range of this group of size 2 is less than the scaled 0.95 quantile of the distribution of the range of a sample of size 2 from normal

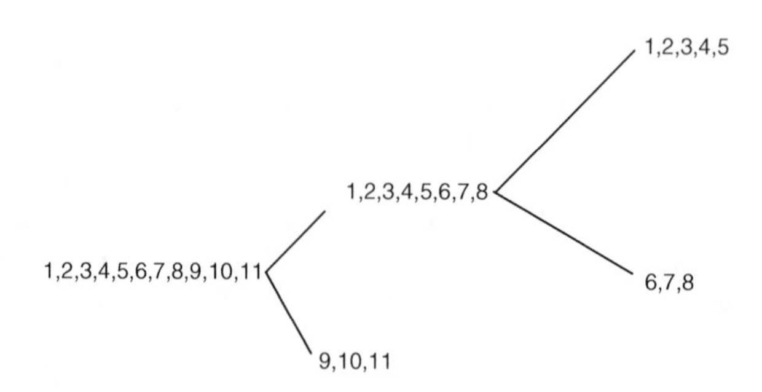

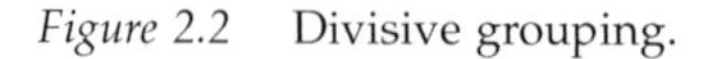

Figure 2.2 Divisive grouping.

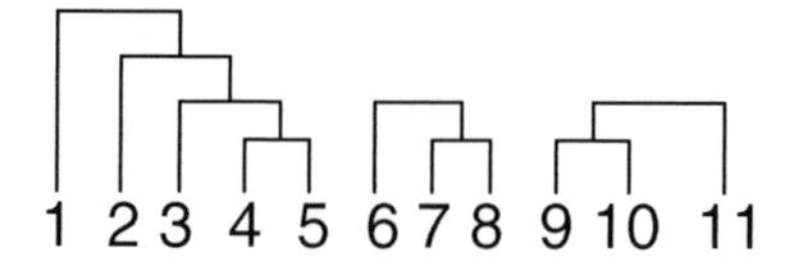

Figure 2.3 Agglomerative grouping.

distribution. Fragment 3 is added next as the range of the group is less than the 0.95 quantile of the distribution of the range of a sample of size 3 from normal distribution, and so on until the procedure attempts to add fragment 6 to the group. The range from fragments 1 to 6 is too big for a sample of size 6, so the procedure is started again on the subgroup formed by fragments 6 to 11. This procedure is repeated until all the fragments have been grouped. Triggs et al.[77,78] show that the divisive approach gives better detection probability on the basis that the divisive approach makes more efficient use of the evidence in the small sample size situation.

In some further research, Curran et al.[76] have shown the impact of using automatic grouping algorithms in a Bayesian interpretation of glass evidence.

2.2.1.3 *Performance*

From the work of Triggs et al.[77,78] there seems little question that the performance of the divisive method (SKM2) exceeds the alternative agglomerative method (EL1). SKM2 also consistently outperforms the modified agglomerative methods (ELM1 and ELM2). The rates of misdetection of groups by SKM2 for very small samples are slightly high, but due to the lack of information in these cases this is not unexpected. These extensive computer simulations, which test the efficacy of the various methods, suggest that SKM2 must be advocated in place of grouping algorithms based on EL1.

An alternative candidate to these methods is subjective grouping "by-eye." It is difficult to make scientific comments on the effectiveness of this method as no tests have been reported and it is difficult to see how they could be performed. They do have the advantage of needing no computing power. While subjectivity is not, per se, wrong, it should be weighed carefully against objective alternatives.

If subjective methods are employed, there appear to be two associated issues that may be discussed.

First is the dot diagram method. If the RIs of the recovered fragments are plotted graphically, this aids "by-eye" grouping. It seems desirable to always use the same scale so that experience hones the "eye."

Second, it seems wise, but not always possible, to cover the control group when grouping. This is easy if there is only one "cluster" of values and has the obvious advantage of removing the possibility of bias. When there are multiple or overlapping clusters, the "blind" grouping might be meaningless. A plausible alternative might be to pursue reasonable "prosecution" groupings and reasonable "defense" groupings as two viable alternatives.

2.2.2 Statistical tests

2.2.2.1 Hypothesis testing

Formal statistical hypothesis testing of data developed at an enormous pace from the 1920s, mainly through the work of R.A. Fisher who established a methodology that rests on the concept of a "null hypothesis." This concept is used in the glass problem, employing the "grouped" approach, in the following way. If the two sets of glass fragments, control and recovered, have come from the same window, then the measurements will all be random variables with the same statistical distribution — this forms the null hypothesis. If we know what sort of statistical distribution is appropriate, then we can use relatively straightforward mathematical methods to predict the behavior of certain functions (called "test statistics") of the x and y. It is necessary to make some assumptions in order for the behavior of the tests to be guaranteed. These assumptions should be kept as explicit as possible. We will discuss two hypothesis tests that are both very similar to each other. These have both been offered in the glass context and implemented in casework. They are the Student's t-test and Welch's modification to the t-test. Both are standard statistical tests and appear in standard texts and statistical packages such as Minitab®, State College, PA; SAS®, Cary, NC; and Microsoft® Excel, Redmond, WA (for instance, in Excel the two-sample equal variance option corresponds to what we will describe here as the t-test and the two sample unequal variance option corresponds to Welch's modification of Student's [Gosset's] original work). Here, we discuss the assumptions made by these tests.

> A1(a): The control and recovered measurements x_1, x_2,..., x_n and y_1, y_2,..., y_m have the same mean; and $\mu_x = \mu_y$ are representative of the distribution of measurements among the fragments which came from the scene window (or broken control object).

This assumption is untested. The recovered and control fragments may be different in terms of their underlying means. In such a case both tests discussed here will be more likely to result in a rejection rate higher than the chosen nominal 1%. Factors that have concerned us include an inference that may also be made from experimental work.[47] Locke and Hayes[47] show that the center of one particular windscreen examined in detail had a different sample mean RI to the edges. It is quite plausible to suggest that in a hit-and-run case the middle of a toughened glass windscreen is left on the road so that only the fragments from the edges can be taken as the control. In such a case $\mu_x \neq \mu_y$. No viable alternative is known at this time. Such an effect should elevate false negatives, but not false positives; therefore, this assumption may, nonetheless, be reliable.

> A1(b): s_x and s_y are estimates of the same parameter σ, the population standard deviation.

This assumption asks whether the two samples both give equally valid estimates of the population parameter σ. It is initially difficult to see why this should not be true, and most early work made this assumption. However, theoretical and experimental considerations suggest that in some cases the recovered group has a larger underlying variance than the control. This may be because they are harder to measure, choice is removed, or because they are "loaded" with surface or near surface fragments. Locke and Hayes[47] observed that most, but not all, of the variance in a window exists across the depth of a single piece.

These authors studied (*inter alia*) the variation in RI across a toughened windscreen. Three areas, A, B, and C, were sampled across the diagonal of the windscreen. Therefore, there are three possible t-test comparisons that can be made between these areas. The data from Locke and Hayes' work are presented in Table 2.1.

The t-values are 2.98 for A vs. B, 4.04 for B vs. C, and 1.41 for A vs. C. Therefore, of the three possible comparisons, two would fail a t-test at the 99% significance level ($t_{48}(0.005) = 2.68$). This suggests that a representative control is necessary. Often in practice a very small control is submitted and even a large control sample may be exclusively from just one part of the window. This problem is likely to be less for untoughened float glass, but

Table 2.1 The Variation Across a Single Windscreen

Area	Number of measurements	Mean	SD
A	25	1.51674	9×10^{-5}
B	25	1.51682	10×10^{-5}
C	25	1.51670	11×10^{-5}

the doubt must still remain as to whether the control submitted is a representative sample.

In a case where a small control is received, it is plausible that a slight underestimate of σ_x is obtained.

We offer two approaches: one that makes assumption A1(b) (Student's *t*-test) and one that does not (Welch's test). It would seem reasonable to make assumption A1(b) and proceed with the *t*-test whenever the samples are of equivalent quality. This would occur, for instance, in a hit and run. Large fragments of glass have been recovered from a roadway and may be compared to similar quality fragments samples from the vehicle or from a ram raid where large pieces of glass are recovered from a vehicle (remember we are discussing the variance effects here not the means). However, in the majority of cases where a clothing sample is compared to "clean" glass from a control sample, we doubt that A1(b) is valid and we recommend proceeding with Welch's test.

> A2: Measurements on glass fragments from a broken window have a normal distribution.

We are not aware of any experimental validation of this assumption and this is a serious omission in glass examination. There are obvious situations where this assumption breaks down, particularly because of surface effects. It is a standard belief that scope for failure of the assumption in a manner which would have a dramatic effect on the test is limited, and this has been confirmed by computer simulations.[81]

> A3: The recovered fragments have all come from the same smashed glass object.

When we make assumption A3, we are assuming that we have first employed some sensible grouping algorithm and are comparing a group of recovered fragments with a group of control fragments. This assumption troubles many experienced examiners with whom it has been discussed. Evett and Lambert[75] argue that the advantages considerably outweigh the disadvantages. They demonstrate the benefits of a grouping/*t*-test combination against a range test approach concentrating on the false positive rate (which, as previously stated, is larger with the range test). A full Bayesian analysis may not require the grouping assumption; however, we are unable to do more than imagine how this might proceed at this time. We believe that at some future time grouping will be seen as an interlude in glass examination. However, at this time the evidence strongly favors it.

2.2.2.1.1 *Student's t-test.* The Student's *t*-test is used to make inferences about a population mean using the information contained in the sample mean. In general, a hypothesis test follows a standard framework.

1. Ask a question.
2. Propose a null hypothesis.
3. Propose a mutually exclusive alternative hypothesis.
4. Collect data.
5. Calculate the test statistic.
6. Compare this test statistic to values that would be expected if the null was true.
7. If the test statistic is "too large," reject the null hypothesis.
8. Answer the original question.

In the context of evaluating forensic glass evidence, the null hypothesis is always the same. Specifically, we hypothesize that the recovered and control glass are two samples from the same window. Mathematically, we are testing the hypothesis H_0: $\mu_x = \mu_y$. An equivalent and more prevalent reformulation of this hypothesis is H_0: $\mu_x - \mu_y = 0$.

If the null hypothesis is true, then it is expected that the difference between the means of the two samples will be small compared to the differences observed between samples from distinct sources. In this case "small" is defined with respect to some measure of the variability of RI within a window (usually the variance or standard deviation).

The comparison of the means of two random samples from distinct normal distributions is treated in many elementary textbooks on statistics. The problem can formally be stated as the following:

We have two independent random samples $x_1, x_2, \ldots, x_n$ and $y_1, y_2, \ldots, y_m$ which are assumed to be independently and identically distributed (iid) as normal random variables. That is $x_i \sim N(\mu_i, \sigma_x^2)$, $i = 1, \ldots, n$ and $y_j \sim N(\mu_y, \sigma_y^2)$, $j = 1, \ldots, m$. Each sample has a sample mean given by

$$\bar{x} = \frac{1}{n}\sum_{i=1}^{n} x_i \text{ and } \bar{y} = \frac{1}{m}\sum_{j=1}^{m} y_j \tag{2.1}$$

respectively, and sample standard deviations given by

$$s_x = \sqrt{\frac{\sum_{i=1}^{n}(x_i - \bar{x})^2}{n-1}} \text{ and } s_y = \sqrt{\frac{\sum_{j=1}^{m}(y_j - \bar{y})^2}{m-1}} \tag{2.2}$$

If the population variances are equal ($\sigma_x = \sigma_y = \sigma$), then the statistic defined by

$$T = \frac{(\bar{x} - \bar{y}) - (\mu_x - \mu_y)}{s_p \sqrt{\frac{1}{n} + \frac{1}{m}}} \tag{2.3}$$

has a t-distribution with $n + m - 2$ degrees of freedom. s_p^2 is a pooled estimate of the common variance σ defined as

$$s_p^2 = \frac{(n-1)s_x^2(m-1)s_y^2}{n+m-2} \tag{2.4}$$

Given the null hypothesis of no difference, the term $\mu_x - \mu_y$ may be removed without effect. It is conventional among users of these tests to set a "significance level" discussed earlier. Table 2.2 shows 1% critical values from a standard t-distribution. These are the values which the test statistic must exceed for the two means to be declared different at the 1% significance level for various degrees of freedom. So if, for example, $n + m - 2 = 10$, then the data shows that, if the null hypothesis is true, there is, on average, only a 1% theoretical chance that the test statistic would exceed 3.169 by random measurement error alone. If the statistic did indeed exceed this value, then the null hypothesis would be rejected at the 1% significance level.

Table 2.2 Table of Critical Values for a t-Test at 1% Significance Level

Degrees of freedom	Critical value for test at 1% significance level
5	4.032
6	3.707
7	3.499
8	3.355
9	3.250
10	3.169
11	3.106
12	3.055
13	3.012
14	2.977
15	2.947
16	2.921

2.2.2.1.2 Welch's modification to the Student's t-test. Welch's modification to Student's t-test[82] uses the same procedure except that the test statistic is slightly different and the degrees of freedom are nonintegral.

As mentioned earlier, the observed variance in a group of glass has two components: the intrinsic variance of the glass fragments and the measurement error.

It seems likely that the measurement error for recovered fragments will be higher than for the control. Recovered fragments may be dirtier and smaller than the fragments freshly broken from the control. This leads to a less distinct disappearance/reappearance of the glass as it is cooled and heated in the oil. The resultant measurement of the RI is less precise than that for a better glass sample. Conversely, for the measurement of the control glass, there is ample glass to mount on a microscope slide, and there is a large number of glass fragment edges to choose from. This usually leads to better measurements on the control glass.

Experiments have been described (Jones and Smalldon, personal communication) where the "choice" of fragment by the operator was removed. In these experiments the variance of the "control" group was increased. They have postulated that this is because of an unconscious choice mechanism that tends to give the same result for consecutive measurements by the operator selecting edges with a similar "look."

Not relevant to this discussion (which is centered on the use of GRIM I and II) but fascinating is the experiment by K.W. Smalldon (personal communication, 1995) where music was played to subjects using the "clicking" Mettler Hot-Stage. Those subjects listening to music had larger variances in their measurements, suggesting that the clicking was interacting in some way with their pressing of the button to record disappearance. Because disappearance and reappearance are visual phenomena, this suggests that the clicking was biasing the results toward a lower variance in some way, perhaps by suggesting when the operator should press the button.

It must also be noted that surface fragments will be overrepresented in the recovered group, and this may also lead to an increase in variability of the recovered group (Jones, K.W. Smalldon, and Crowley, personal communication).

The control sample submitted is not always a good sample. It often contains only one or a few small pieces of glass, and these cannot be expected to represent the full range of variability present in the window or windscreen. For example, Locke and Hayes[47] measured the RIs across the diagonal of a windscreen. The variance across the whole windscreen was higher than that at any one point.

Therefore, since it is unlikely that the control sample is always truly a random sample, the standard deviation of the control sample will underestimate the actual standard deviation or variance of the window.

A credible line of argument can be made that a larger variance for recovered glass must be expected. We have previously mentioned that tests (Reference 37 and C.A. Coulson, J.M. Curran, and A.B. Gummer, personal communication, 1995) suggest that there is an increase in the variation of measurement with lower edge count, but that the mean remains approximately the same. Since using GRIM, it is easily possible to get 99, 99 edge counts for control fragments, whereas recovered fragments are typically less. It follows that the variance of the recovered group must be larger.

Whatever the explanation, and there are many contenders, it is an empirical fact that the standard deviation of the recovered group is typically larger, often 50% larger, than the control. Pinchin and Buckleton (personal communication) collected data from three Forensic Science Service (FSS) laboratories and observed that the variance was, in fact, typically larger for the recovered group. A plot of the standard deviation of the recovered groups against the control for this data is given in Figure 2.4. It can be seen that most of the points lie above the $x = y$ line.

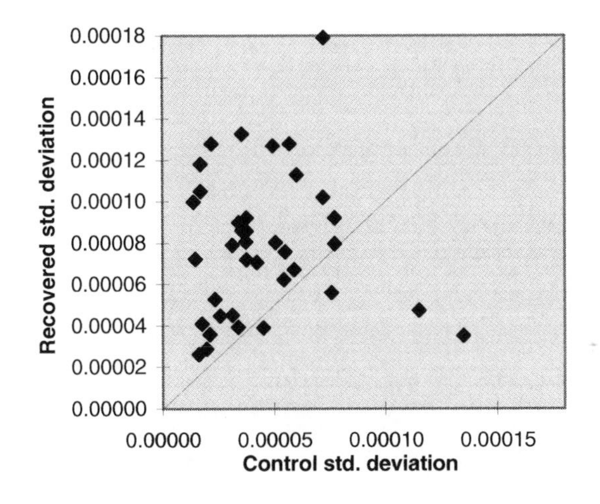

Figure 2.4 Standard deviation of recovered groups against controls for data from R.A. Pinchin and J.S. Buckleton, 1993.

We have experienced situations where the control group fragments seem very close in RI, giving a very small estimate for the standard deviation of this group. One possible result of this is that the samples may fail the t-test even though the groups are very close in absolute RI.

If the control and recovered groups are from different pieces of glass then well and good, but if it is an artifact of the higher variability of the recovered group and the low variability of the control group then the result is unacceptable and a remedy is required. We point out that this is a minor point in the context of glass examination and has been described as a mindless detail. It can easily and justifiably be argued that the t-test would outperform range tests by an even larger margin if the variance of the recovered samples was high. We proceed here, however, to offer an option to those glass examiners who are concerned about some of these details.

Often the size of the recovered sample is small, possibly as small as 2 or 3 fragments, while the control sample can be arbitrarily large. Usual practice takes a control sample of 6 to 10 fragments. In such cases the estimate of the pooled standard deviation is dominated by the larger sample in the t-test procedure and the larger variance of the recovered fragments goes largely unnoticed.

To overcome the problem of false discrimination because of poor variance estimates, we advocate the use of Welch's modification[82] to the t-test. This makes allowance for the variance of the recovered group and the control group to be unequal. This means that assumption A1(b) is not made; however, the remaining assumptions are.

Welch's statistic is given by

$$V = \frac{(\bar{x} - \bar{y}) - (\mu_x - \mu_y)}{\sqrt{\frac{s_x^2}{n} + \frac{s_y^2}{m}}} \tag{2.5}$$

Again, assumption A1(a) simplifies this expression to

$$V = \frac{(\bar{x} - \bar{y})}{\sqrt{\frac{s_x^2}{n} + \frac{s_y^2}{m}}} \tag{2.6}$$

The distribution of V is no longer exactly t, but may be approximated by a t_ψ distribution whose degrees of freedom, ψ, may be estimated from the data and are

$$\psi = \frac{\left(\frac{s_x^2}{n} + \frac{s_y^2}{m}\right)}{\left(\frac{s_x^4}{n^2(n-1)} + \frac{s_y^4}{m^2(m-1)}\right)} \tag{2.7}$$

which is typically nonintegral. This modified statistic is used as the default to carry out two-sample t-tests in such widely available statistical packages as SAS and exists as an option in Microsoft Excel.

Percentage points for t-distributions with nonintegral degrees of freedom can be quickly and accurately computed.[83] If we take $[\psi + 1]$, the integral part of $\psi + 1$, for the degrees of freedom, the percentage points can be obtained from standard tables of the t-distribution. Using $[\psi + 1]$ will provide slightly smaller P-values than using the exact degrees of freedom, thus leading to conservative test procedures. Many studies have shown that the performance of V in practice is entirely satisfactory.[84,85]

2.2.2.2 *How many control fragments?*

The practical consequences of examining the behavior of the testing procedures are that if the number of recovered samples is small, there is no benefit in estimating the RI of a large number of control samples and there may be some harm.

Since the size of the recovered sample, m, is often beyond the control of the investigator, whereas the size of the control sample, n, can be made very large, use of the t-statistic will invariably lead to domination of the estimate of the standard error of the difference of the means by the standard deviation of the control sample. This is untenable for two reasons. The first, discussed earlier, is the underestimation of the standard error. The second arises from the number of degrees of freedom (d.f.) of the t-test, $n + m - 2$, which will apparently cause the test to become more sensitive as n increases.

The behavior of t and V in the limit where m, the size of the recovered sample, is small and n, the size of the control sample, is very large is instructive.

$$T \cong \frac{\bar{x} - \bar{y}}{s_x / \sqrt{m}} \text{ with } n + m - 2 \text{ d.f. and } V \cong \frac{\bar{x} - \bar{y}}{s_y / \sqrt{m}} \text{ with } m - 1 \text{ d.f.} \quad (2.8)$$

Using T to test the hypothesis that $\mu_x = \mu_y$ is tantamount to testing whether $\bar{y}$ can be regarded as coming from a normal distribution with mean $\bar{x}$ and a variance σ_x^2. On the other hand, the use of V tests whether $\bar{y}$ comes from a normal distribution with mean $\bar{x}$ and a variance σ_y^2. Since we expect that in practice $\sigma_y^2 > \sigma_x^2$, and since V will be tested against a t-distribution with many fewer degrees of freedom, routine use of the t-statistic will lead to oversensitive tests.

Even if Welch's test is used, there appears to be a "law of diminishing returns." Increasing the number of control measurements should improve the estimates of σ_x^2 and $\bar{x}$. However, there is not an equivalent improvement in the estimation of

$$\sqrt{\frac{s_x^2}{n} + \frac{s_y^2}{m}} \quad (2.9)$$

because as n becomes large this is dominated by s_y^2/m which is fixed by the recovered sample obtained. This suggests that the number of control fragments can be set relative to the number of recovered fragments obtained (see Table 2.3).

Table 2.3 A Plausible Number of Controls to Take Conditional on the Size of the Recovered Group

Number of recovered fragments (m)	Recommended number of controls (n)
1	4
2	4
3	6
4	6
5	8
6	9
7	10
8	10
9	10
10	10
12	10
15	10
20	10

We have investigated the performance of these combinations relative to the nominal α level by simulation; however, we are still investigating the logic of these simulations and do not present them here.

2.2.2.3 *Setting significance levels*

If we assume that a significance level is chosen, say α, this may be interpreted as the theoretical false exclusion rate. This false exclusion rate is theoretical in that it depends on a number of assumptions being met, which, to a greater or lesser extent, may be violated as discussed previously. In such a case the empirical false exclusion may be more or less than 1%. However, let us proceed with the concept that the true false exclusion rate is α. It is necessary to set α in order to proceed with the concept of hypothesis testing.

A compelling argument might be to set α such that the false exclusion rate α plus the false inclusion rate β is minimized. Since β is case dependent, this would have to be performed on a per case basis. This is, nonetheless, the logical conclusion of the hypothesis testing approach or, alternatively, the first faltering steps to the Bayesian approach.

Some countries have a legal maxim, "it is better that 20 guilty men go free than to convict one innocent man." Can this be used? At face value this might suggest that a viable procedure would be to set $\alpha = 20\ \beta$. This holds if, and only if, matching and nonmatching groups occur in equal proportions (this follows from a chain of logic more prevalently discussed in the field of DNA evidence called the fallacy of the transposed conditional or the prosecutors which we will explain later). At face value then, if we wish to set the false exclusion rate to 20 times the false inclusion rate and if the probability of the group matching or not before measurement of the RIs is equal, then we set $\alpha = 20\ \beta$. This is not normal forensic practice.

2.2.2.4 *Elemental composition measurements — Hotelling's T^2*

Researchers have found that elemental composition comparisons add discrimination potential to distinguish between glass fragments when the RI does not.[41,58,66] The elements of interest are the minor and trace elements: aluminium (Al), iron (Fe), magnesium (Mg), manganese (Mn), strontium (Sr), zirconium (Zr), calcium (Ca), barium (Ba), and titanium (Ti). Trace elements such as strontium and zirconium are present in the low parts per million concentration range, and preliminary work on small data sets has shown that these elements have little or no observable correlation between the other elements,[41,58,66,86,87] which makes them very good discriminating "probes."

Traditional treatment of the data involves determining the mean concentration and the standard deviation for each element and then comparing the means using a "3 sigma rule" and testing the match criteria to determine if the ranges overlap for all of the elements. If any of the elements fails this test, then the fragments are considered not to match.

The following section will demonstrate a statistical test that has advantages over the 3 sigma rule approach.

2.2.2.4.1 The multiple comparison problem. The 3 sigma rule has two problems. The first is the problem of multiple comparisons. The 3 sigma rule has a theoretical false rejection rate of approximately 0.1%, i.e., on average one time in one thousand the scientist will say the mean concentrations are different when, in fact, they are the same. This empirical false rejection rate is much higher when small numbers of fragments are being compared. Each comparison for each element has the same rate of false rejection; however, the overall rate is much larger, even if the elemental concentrations are independent. Consider the following: I have an extremely biased coin with the probability of getting a tail equal to $\Pr(T) = p = 0.01$. The outcome of each coin toss is independent of any previous toss. If the number of tosses, n, is fixed and X is the random variable that records the number of tails observed, then X is binomially distributed with parameters n and p. This experiment is analogous to making pairwise comparisons on n element concentrations. If $n = 10$ and the probability of a false rejection on one element is 0.01, then the probability that at least one false rejection will be made is 0.096 or 9.6%, so the overall false rejection rate is nearly ten times higher than the desired rate. In general, if the false rejection rate, or size, of a procedure is α for a single comparison, and n comparisons are performed in total, then the overall size is approximately $n \times \alpha$. A simple solution is to increase the width of the intervals so that the size of the individual comparison is α/n. This is known as the Bonferroni correction and its immediate drawback is obvious — as n, the number of elements, increases, it becomes almost impossible to detect any difference between the two means for any given element.

The second problem with the 3 sigma rule approach is that is fails to take into account any estimated correlation between the elements. That is, the estimated concentration of one element will be associated with the estimated concentration of another element. Failure to include this information results in severe underestimation of any joint probability calculation.

The solution to both these problems is the multivariate analog to the Student's *t*-test.

2.2.2.4.2 *Hotelling's T^2 — A method for comparing two multivariate mean vectors.* The Student's *t*-test and Welch's modification have been used and discussed extensively in the treatment of RI measurements.[8,75] Comparison of glass samples with respect to RIs examines the standardized distance between the two sample means. Hotelling's T^2 (named after Harold Hotelling, the first statistician to obtain the distribution of the T^2 statistic) is a multivariate analog of the *t*-test that examines the standardized squared distance between two points in *p*-dimensional space.[88] These two points, of course, are given by the estimated mean concentration of the discriminating elements in both samples.

Suppose that m fragments have been recovered from the suspect, n control fragments have been selected from a crime scene sample, and $n + m > p + 1$, where p is the number of elements considered, then T^2 has a scaled *F*-distribution

$$T^2 \sim \frac{(n+m-2)p}{(n+m-p-1)} F_{p,n+m-p-1} \tag{2.10}$$

Use of the *F*-distribution depends on two assumptions about the statistical distribution of the data: (1) both samples come from a multivariate normal distribution and (2) both populations have the same covariance structure,[89] i.e., the spread of the elemental concentration is approximately the same in each. Large values of T^2 are evidence against the hypothesis of no difference between the two populations, i.e., evidence against a match.[88]

2.2.2.4.3 *Examples.* The data in the following examples come from two distinct sources, one brown bottle and one colorless bottle taken from a different process line at the same plant at the same time. Ten fragments were taken from each bottle, and the concentrations of aluminum, calcium, barium, iron, and magnesium ($p = 5$) were determined by ICP-AES.

The first example uses five fragments from the brown bottle as a control sample ($n = 5$) and five fragments from the same bottle as a recovered sample ($m = 5$) so that the population means are truly equal. Hotelling's $T^2 = 11.69$ and $F_{5,4}\,(0.01) = 15.521$, so

$$T^2 = 11.69 << \frac{(5+5-2)5}{(5+5-5-1)} F_{5,4}(0.01) = 10F_{5,4}(0.01) = 155.21 \tag{2.11}$$

T^2 is comparatively small in relation to F, thus there is no evidence to reject the null hypothesis, i.e., there is no evidence to suggest that the two samples come from different sources.

The second example takes ten fragments from the brown bottle as the control sample ($n = 10$) and ten fragments from the colorless bottle as the recovered sample ($m = 10$), so the null hypothesis is false, i.e., the population means are truly different. Hotelling's $T^2 = 708.86$ and $F_{5,14}$ (0.01) = 4.69, so

$$T^2 = 708.86 >> \frac{(10+10-2)5}{(10+10-5-1)} F_{5,14}(0.01) = \frac{90}{14} F_{5,14}(0.01) = 30.15 \quad (2.12)$$

In this example T^2 is comparatively large in relation to F, thus there is very strong evidence to reject the null hypothesis, i.e., there is very strong evidence to suggest that the two samples come from different sources.

2.2.2.4.4 Discussion on the use of Hotelling's T^2. Hotelling's T^2 test for the difference in two mean vectors provides a valid statistical method for the discrimination between two samples of glass based on elemental data. The user may decide whether or not to include an RI measurement comparison, but the test remains the same. The properties of the test are closer to the desired properties of the 3 sigma rule than the rule itself. Hotelling's T^2 must be advocated in place of any algorithms based on the 3 sigma rule or modifications of it.

In general, however, it would be desirable to abandon tests altogether and move toward a direct calculation of a likelihood ratio from continuous multivariate data.

2.3 *Coincidence probabilities*

Consider a case in which there is one group of recovered fragments which is indistinguishable from the control on the basis of the t-test at the 1% level and on $n + m - 2$ degrees of freedom. A conclusion that "the recovered glass could have come from the same source as the control" is inadequate. Indeed, it is little more than a statement of the obvious. If we are to be balanced in our approach then we must add "but it could also have come from some other source." In order to assist the deliberation of a court of law, it is necessary that we give some indication of the strength of the evidence.

The traditional forensic approach to this kind of problem is to invoke the idea of "coincidence." If the recovered fragments have not come from the scene window then the match is a coincidence. What is the chance of such a coincidence occurring? This leads to the question "if I take control fragments from some glass source other than the crime window, what is the chance they would match the recovered, using the same comparison test?" Intuition tells us that the smaller this chance is, the greater the evidential value of the observed match. Evett and Lambert[75] were among the first to specifically define a coincidence probability. "The coincidence probability … is the probability that a set of m fragments taken at random from some source in the population ... would be found to be similar .... with n control measurements of mean …." It should be noted that the definition given by Evett and Lambert and the one referred to as desirable later in this book are similar, but not the same. The former refers to fragments that would match the control, and the latter refers to fragments that would match the recovered. This difference was elegantly pointed out by Stoney.[90] Deeper reflection on the question has followed, championed in the forensic community by Evett, and is continued in Chapter 3.

Following a precise definition of what is meant by "coincidence probability," it is easy to develop an algorithm that can estimate this probability.

We are unable, however, to give a precise (and correct) definition of coincidence probability. Later, we will want the probability of this evidence (the recovered fragments) if they have not come from the control. Since we are assessing the probability of the evidence if the glass did not come from the control window, it is difficult to see how the n measurements of the control have anything to do with the definition.

However, in an attempt to proceed, we define this probability as an answer to the following question. "What is the probability that a group of m fragments taken from the clothing of a person also taken at random from the population of persons unconnected with this crime would have a mean RI within a match window of the recovered mean?" Defining the match window is also difficult, and most attempts are based on applying the match criterion to m fragments from random clothing (or other population) and the n fragments from the control. Such a match window would be centered on the recovered mean rather than the control mean, as implied in the question given by Evett, but of the same width as that implied by Evett.

Common sense tells us that the smaller the coincidence probability is, the greater the evidential value of the observed match.[91] The coincidence probability may be interpreted as the probability of false inclusion, β, referred to earlier.

Although this coincidence probability approach has strong intuitive appeal, it has many problems. What do we mean by "some glass source other than the crime window?" Do we mean another *window*? What other kinds of glass sources should we consider?

The more we think about these questions, the more we are led to suspect that our original question in relation to coincidence may not be the best one. Our practices in this area have been conditioned largely by two factors. The first is that it has always been easier to use as a source of data the control glass samples that are submitted during routine casework. The second factor is the availability of elemental analysis, which in many cases can establish whether or not the recovered fragments are window glass.

Another problem arises when, for example, we have two groups of recovered fragments — one matches the control and the other does not.

When we try to formulate our coincidence probability we soon run into logical inconsistencies which have never been fully resolved and which were discussed in later work by Evett and Lambert.[92]

Whatever we do with the coincidence approach we must recognize that we are taking a fairly narrow look at the evidence.[93,94] Intuitively, we will recognize that there are other considerations which may be every bit as important, particularly those of transfer and persistence. Although research has been done which may help us assess the likelihood of transfer and persistence in a given case, we need some kind of logical framework to roll that assessment into the overall appreciation of evidential strength. The coincidence approach has no means of doing this.

2.4 Summary

The conventional approach to glass evidence described here is essentially a two-stage approach: matching and coincidence estimation. As we have seen, there are a number of problems.

1. The principles of this kind of approach have never been clearly established, and the approach does not have a clear, logical framework.
2. The matching stage is open to challenge if it cannot be seen to embody clear statistical principles. Yet significance tests offer an air of objectivity that is illusory because they inevitably require assumptions, the validity of which in any particular case involves human judgment.
3. The choice of significance level for the matching test is entirely arbitrary, and the sharp distinction between match and nonmatch does not make much sense.
4. The coincidence approach to estimating evidential weight does not lead to the formulation of pertinent questions when considering the defense perspective. In particular, it is incapable of dealing with problems of multiple controls and/or recovered groups.
5. The two-stage approach provides no framework for taking account of the considerations of transfer and persistence in forming an overall assessment of the evidence.

In order to interpret glass evidence, we require an infrastructure that enables us to do the following.

1. Identify the questions which are not only relevant to the needs of the criminal justice system, but are also within our capabilities to answer.
2. Work within a logical framework based on scientific principles.
3. Utilize the most useful statistical methods whenever they are relevant, but subject to the disciplined use of expert judgment.
4. Design and create databases.

5. Utilize the available databases effectively within the context of expert judgment.
6. Identify and explore all of the hypotheses which might be relevant in an individual case, giving them weight in a balanced and scientific manner.

As we will see in the next chapter, using the Bayesian framework allows us to satisfy most of these criteria.

2.5 Appendix A

Scott Knott Modification 2 (SKM2)

1. Sort $y_1, y_2, \ldots, y_m$ into ascending order and label the sorted observations $y_{(1)}, y_{(2)}, \ldots, y_{(m)}$.
2. For $j = 1$ to $m - 1$, calculate the following

$$B_j = j\left(\bar{y}_1 - \bar{\bar{y}}\right)^2 + (k - j)\left(\bar{y}_2 - \bar{\bar{y}}\right)^2 \quad (2.13)$$

   where $\bar{y}_1$ is the mean of the fragments $y_{(1)}$ to $y_{(j)}$, $\bar{y}_2$ is the mean of the fragments $y_{(j+1)}$ to $y_{(m)}$, and $\bar{\bar{y}}$ is the mean of all the fragments, i.e., calculate the BGSS for each of the $m - 1$ ordered partitions of the data into two subgroups.
3. Find the maximum BGSS B_0 and record the position where it was found (say j^*).
4. Calculate the statistic proposed by Scott and Knott,[80]

$$\lambda_* = \frac{\pi}{2(\pi - 2)} \frac{B_0}{s_0^2} \quad (2.14)$$

5. If $\lambda_* > \lambda_*(\alpha;m)$, then split the fragments $y_{(1)}, y_{(2)}, \ldots, y_{(m)}$ into two groups, $\{y_{(1)}, y_{(2)}, \ldots, y_{(j^*)}\}$ and $\{y_{(j^*+1)}, y_{(2)}, \ldots, y_{(m)}\}$.
6. If there was a split in Step 4, repeat Steps 2 through 5 for each new subgroup until no more splits can be made.

Evett Lambert 1 (EL1)

1. Sort $y_1, y_2, \ldots, y_m$ into ascending order and label the sorted observations $y_{(1)}, y_{(2)}, \ldots, y_{(m)}$.
2. Select $x_{(M)}$, where M is the greatest integer less than or equal to $^1/_2\,(m + 1)$, e.g., if $m = 11$, $M = 6$.

3. Find the next nearest fragment to $y_{(M)}$, which will be $y_{(M-1)}$ or $y_{(M+1)}$, and combine it with $y_{(M)}$ to form a group of 2. If the range of the group is within the required limits, find the next nearest fragment and compare the range for a group of size 3, and so on. The critical values $r(\alpha;m)$ are listed in Table 2.4 in Appendix B.
4. Repeat Step 3 until all the elements are in one group, or go to Step 2 for any subgroups of size >1.

Evett Lambert Modification 1 (ELM1)

1. Sort $y_1, y_2,\ldots, y_m$ into ascending order and label the sorted observations $y_{(1)}, y_{(2)},\ldots, y_{(m)}$.
2. Check that the range of $y_{(1)}, y_{(2)},\ldots, y_{(m)}$, $r = y_{(m)} - y_{(1)}$ is less than or equal to the 100(1–α)% critical value in Table 2.5 in Appendix B, $r(\alpha;m)$. If $r \le r(\alpha;m)$, then the fragments are considered to have come from only one source, and, therefore, there is no need to go any further. If this is not the case, i.e., $r > r(\alpha;m)$, then go to Step 3.
3. Select $y_{(M)}$, where M is the greatest integer less than or equal to $^1/_2$ $(m+ 1)$, e.g., $m = 11$, $M = 6$.
4. Find the next nearest fragment to $y_{(M)}$, which will be $y_{(M-1)}$ or $y_{(M+1)}$, and combine it with $y_{(M)}$ to form a group of 2. If the range of the group, r, is within the required limits, find the next nearest fragment and compare the range for a group of size 3, and so on. The required limits are listed in Table 2.5 in Appendix B.
5. Repeat Step 3 until all the elements are in one group, or go to Step 2 for any subgroups of size >1.

Evett and Lambert Modification 2 (ELM2)

1. Sort $y_1, y_2,\ldots, y_m$ into ascending order and label the sorted observations $y_{(1)}, y_{(2)},\ldots, y_{(m)}$.
2. Check that the range of $y_{(1)}, y_{(2)},\ldots, y_{(m)}$, $r = y_{(m)} - y_{(1)}$ is less than or equal to the 100(1 – α)% critical value in Table 2.6 in Appendix B, $r(\alpha;m)$. If $r \le r(\alpha;m)$, then the fragments are considered to have come from only one source, and, therefore, there is no need to go any further. If this is not the case, i.e., $r > r(\alpha;m)$, then go to Step 3.
3. Find the smallest gap between the fragments, i.e., find i and j such that $|y_{(i)} - y_{(j)}|$ is minimized for $1 \le j < i \le m$.
4. If $|y_{(i)} - y_{(j)}| \le r(\alpha;2)$, i.e., if the range of the group consisting of fragments $y_{(i)}$ and $y_{(j)}$ is less than the range of a group of size 2 from a normal distribution, find the nearest neighbor to $y_{(i)}$ or $y_{(j)}$ and compare the range against $r(\alpha;3)$, etc.
5. Repeat Step 4 until all the elements are in one group, or go to Step 2 for any subgroups of size >1.

2.6 Appendix B

Table 2.4 90, 95, and 99% Critical Values for λ_* Given m

m	$\lambda_*(0.01;m)$	$\lambda_*(0.05;m)$	$\lambda_*(0.01;m)$
2	3.73	5.33	9.29
3	5.74	7.38	11.81
4	7.39	9.56	13.46
5	8.88	10.91	15.21
6	10.23	12.32	16.94
7	11.47	13.61	18.29
8	12.81	15.23	20.64
9	13.83	16.37	21.82
10	15.18	17.54	23.09
11	16.42	18.89	24.20
12	17.42	19.84	25.44
13	18.79	1.24	27.03
14	19.78	22.49	28.02
15	20.89	23.77	29.84
16	22.13	24.80	31.40
17	22.91	25.53	31.61
18	24.25	27.03	32.66
19	25.42	28.31	34.05
20	26.45	29.20	35.45

Table 2.5 95% Critical Values for the Range of a Sample of Size m from a Normal Distribution Scaled by a Weighted Estimate of Standard Deviation[75]

m	10,000 × 95th percentile
2	1.13
3	1.46
4	1.68
5	1.86
6	1.98
7	2.09
8	2.21
9	2.29
10	2.36
11	2.44
12	2.50
13	2.56
14	2.61
15	2.66
16	2.70
17	2.74
18	2.78
19	2.82
20	2.85

Table 2.6 90, 95, and 99% Critical Values for the Range of a Sample of Size m from a $N(0,s^2)$ Distribution with $s = 4 \times 10^{-5}$

m	90th percentile $r(0.1;m)$	95th percentile $r(0,05;m)$	99th percentile $r(0.01;m)$
2	0.93	1.11	1.47
3	1.16	1.32	1.64
4	1.28	1.44	1.76
5	1.38	1.54	1.82
6	1.46	1.60	1.89
7	1.52	1.65	1.94
8	1.57	1.71	2.01
9	1.62	1.76	2.06
10	1.65	1.80	2.05
11	1.70	1.83	2.09
12	1.71	1.84	2.13
13	1.74	1.88	2.14
14	1.77	1.88	2.15
15	1.79	1.92	2.16
16	1.80	1.94	2.19
17	1.81	1.94	2.19
18	1.85	1.99	2.25
19	1.86	1.98	2.20
20	1.87	2.00	2.27

2.7 *Appendix C*[88]

Suppose that $x_i = [x_{i1},\ldots, x_{ip}]^T$ are the elemental concentrations of p elements on the ith control fragment, and $y_j = [y_{j1},\ldots, y_{jp}]^T$ are the elemental concentrations of p elements on the jth recovered fragment. If n control fragments are to be compared with m recovered fragments, the matrices

$$\begin{bmatrix} x_{11} \cdots x_{np} \\ x_{12} \cdots x_{np} \\ \vdots \quad \cdots \quad \vdots \\ x_{1p} \cdots x_{np} \end{bmatrix}, \begin{bmatrix} y_{11} \cdots y_{mp} \\ y_{12} \cdots y_{mp} \\ \vdots \quad \cdots \quad \vdots \\ y_{1p} \cdots y_{mp} \end{bmatrix} \tag{2.15}$$

represent the measurements. The summary statistics are defined by the mean vectors

$$\bar{x} = \frac{1}{n}\sum_{i=1}^{n} x_i \text{ and } \bar{y} = \frac{1}{m}\sum_{j=1}^{m} y_j \tag{2.16}$$

and the sample covariance matrices.

$$S_X = \frac{1}{n-1}\sum_{j=1}^{n}\left(x_j - \bar{x}\right)\left(x_j - \bar{x}\right)^T \text{ and } S_Y = \frac{1}{m-1}\sum_{j=1}^{m}\left(y_j - \bar{y}\right)\left(y_j - \bar{y}\right)^T \quad (2.17)$$

respectively. An estimate of the common covariance matrix, Σ, is given by

$$S_{pooled} = \frac{(n-1)S_X + (m-1)S_Y}{n+m-2} \quad (2.18)$$

Hotelling's T^2 is then defined as

$$T^2 = \left(\bar{x} - \bar{y}\right)\left[\left(\tfrac{1}{n} + \tfrac{1}{m}\right)S_{pooled}\right]^{-1}\left(\bar{x} - \bar{y}\right)^T \quad (2.19)$$

chapter three

The Bayesian approach to evidence interpretation

The most important task of the forensic scientist is to interpret evidence. As we have seen, the conventional way to approach evidence does not allow one to take into consideration parameters such as the presence of glass, transfer, and persistence. Nor does it allow one to assess correctly the value of glass when there are multiple control and recovered groups. Therefore, several authors[5-7,94-98] have indicated that Bayesian inference is the appropriate approach. Indeed, this approach allows the expert not only to address the right questions to assist the court and elaborate research projects focused on specific parameters, but it also allows the expert to avoid pitfalls of intuition.

The application of this thinking to the interpretation of glass evidence was anticipated in other forensic and legal fields at a very early stage.[99]

Robertson and Vignaux[94] draw our attention to an interesting example. Suppose a jury is required to determine whether or not a child has been abused. The jury is told by an expert that the child rocks and that only 3% of nonabused children rock. The jury might be inclined to conclude that because the "coincidence" probability (the probability that the child rocks if he/she is not abused) is small it is safe to infer that the alternative (the child is abused) is true.

Clearly, the jurors' views would change if they were then told that 3% of abused children also rock. The evidence is clearly equally probable whether the child was abused or not and, therefore, has no evidential value.

This small thought experiment leads quickly to the realization that it is the ratio of the probability of the evidence under each of two (or more) hypotheses that defines evidential value.

More formal mathematics also leads to this conclusion, and this result was given by the Reverend Thomas Bayes in 1763.[100] Before stating Bayes Theorem, it is necessary to make some definitions.

3.1 Probability — some definitions

- *Random experiment* — an experiment in which the set of possible outcomes is known, but the outcome for a particular experiment is unknown. For example, tossing a fair coin. The outcomes are heads and tails, but exactly which will come up on any one throw is unknown.
- *Outcome* — the result of a random experiment. In the coin tossing example the outcomes are heads or tails. If the experiment is observing the value of a card drawn out of a deck, the outcome may be any one of {2, 3, 4, 5, 6, 7, 8, 9, 10, J, Q, K, A}.
- *Sample space* — the set of all possible outcomes. The sample space is sometimes denoted by Ω.
- *Event* — a set of one or more outcomes. For example, one might ask about the event of drawing a 10 or higher from a deck of cards. This event, E, is defined by the set $E = \{10, J, Q, K, A\}$.
- *Event occurring* — an event is said to have "occurred" if the result of a random experiment is one or more of the outcomes defined in the event. For example, if a single card is drawn and it is a Jack, then the event E given above is said to have occurred.
- *Equiprobable or equally likely outcomes* — outcomes of a random experiment are called "equiprobable" or "equally likely" if they have the same chance of occurring.

Standard statistical texts often define the probability of an event two ways, both closely related.

Definition

If all outcomes from a random experiment are equiprobable, then the probability of an event E, $\Pr(E)$, is given by

$$\Pr(E) = \frac{\#\ \text{Outcomes in } E}{\text{Total}\ \#\ \text{of outcomes}} \tag{3.1}$$

Definition

The probability of event E is given by the long-term frequency that event E occurs in a very large number of random experiments all conducted in the same way.

These two definitions can be demonstrated by the coin tossing example. Assume one is interested in the probability of a getting a head in a single toss. If the coin is fair, then the probability of getting a head is equal to the probability of getting a tail, i.e., the outcomes are equally likely. So if E = "a head" = {H}, and the set of all outcomes is $\Omega = \{H, T\}$, then

$$\Pr(E) = \frac{\text{\# Outcomes in } E}{\text{Total \# of outcomes}} = \frac{1}{2} = 0.5 \tag{3.2}$$

as we would expect. Similarly, we could repeat our experiment many times and observe the frequency with which event E occurs, i.e., how many heads do we see.

Table 3.1 A Coin Tossing Experiment

# Tosses	# Heads in *n* tosses	Frequency
10	6	0.60000
100	52	0.52000
1000	463	0.46300
10,000	4908	0.49080
100,000	49,573	0.49573

Table 3.1 shows a simulated coin tossing experiment. After 10 tosses, the experimenters had obtained 6 heads, so they estimate the probability of a head by the frequency, i.e., $\Pr(E) \approx 0.6$. After 100 tosses, the experimenters had seen 52 heads, so $\Pr(E) \approx 0.52$. If the experimenters could continue this experiment, i.e., for an infinite number of tosses, they would find $\Pr(E) = 0.5$.

Most readers will be familiar with one or other of these *frequentist* or *classical* definitions of probability. While these definitions form the basis of modern statistics, they are unsatisfactory for the interpretation of scientific evidence.

Definition

A probability is a rational measure of the degree of belief in the truth of an assertion based on information.[94]

This second definition is a *Bayesian* or *subjectivist* definition of probability and is one of the underlying concepts in the whole of Bayesian philosophy. It should be apparent that this definition is more suited to our purposes. For the purpose of illustration, take the question "what is the probability that the suspect committed the crime?" There are two possible outcomes in the "experiment" — either the suspect committed the crime or he did not. However, we have no reason to believe that these outcomes are equally likely. Similarly, the suspect may be a first-time offender, so a long-term frequency approach will not work. The only way this question can be answered is with the aid of evidence. We can also examine the coin tossing example in a Bayesian framework. Realistically, we cannot toss the coin forever, and so any long-term frequency is merely an approximation of the true probability. At some point, however, our approximation will become sufficiently accurate for us to believe that true probability of a head is a half. That is, the evidence is sufficiently strong for us to form a belief about the probability of observing a head. For a more thorough discussion on these concepts the reader is advised to try either Robertson and Vignaux,[94] Bernardo and Smith,[101] or Evett and Weir.[102]

3.2 *The laws of probability*

There are three laws of probability. These laws hold regardless of whether one subscribes to the frequentist or Bayesian definition of probability. We define these laws for two events, but it is simple to generalize the results to more than two events.

3.2.1 *The first law of probability*

The first law of probability is the simplest. It says that all probabilities lie between zero and one. That is no probability can be less than zero or greater than one. Mathematically, the first law of probability is

$$0 \leq \Pr(A) \leq 1 \text{ for any event } A \tag{3.3}$$

As a corollary to this, it is useful to note that the probability of any event A that consists of no outcomes, i.e., the empty set, is zero ($\Pr(A = \phi) = 0$). If the event A consists of all the outcomes, i.e., the sample space, then the probability of A is one ($\Pr(\Omega) = 1$). If A consists entirely of outcomes that are not in the sample space, then the probability of A is zero.

3.2.2 *The second law of probability*

Definition

Two events A and B are said to be *mutually exclusive* if the occurrence of one event precludes the other from happening.

For example, let A be the event that the coin comes up heads, and B be the event that the coin comes up tails, then A and B are mutually exclusive events.

Definition

The second law of probability: if A and B are mutually exclusive events, then the probability of A or B occurring is

$$\Pr(A \text{ or } B) = \Pr(A) + \Pr(B) \tag{3.4}$$

Sometimes $\Pr(A \text{ or } B)$ is written as the probability of A union B, $\Pr(A \cup B)$.

Corollary

If A and B are *mutually exhaustive*, i.e., the outcomes in A and B make up the set of all possible outcomes ($A \cup B = \Omega$), then $\Pr(A \text{ or } B) = 1$.

Proof

$$\Pr(A \text{ or } B) = \Pr(A \cup B) = \Pr(\Omega) = 1$$

If two (and only two) events A and B are mutually exhaustive, then A and B are said to be complementary, and we write B as $\bar{A}$.

3.2.3 *The third law of probability*

Before we can state the third law of probability, it is necessary to introduce the concept of conditional probability. A conditional probability gives the probability of a particular event *conditional* on the fact that some prior event, say B, has occurred. This conditioning may have no effect at all or a very large effect. We write this conditional probability as $\Pr(A \mid B)$ and read it as the probability of A *given* B. All probabilities are conditional. For example, in our coin tossing experiment, when we evaluated the probability of A given B, we conditioned the probability on the fact that we knew the coin was fair.

Definition

The third law of probability: the *joint* probability of two events A and B is given by

$$\Pr(A \text{ and } B) = \Pr(A \mid B)\Pr(B) = \Pr(B \mid A)\Pr(A) = \Pr(B \text{ and } A) \tag{3.5}$$

Two important concepts arise out of the third law of probability: those of *statistical independence* and *conditional independence.*

Definition

Two events A and B are said to be *statistically independent* if, and only if,

$$\Pr(A \text{ and } B) = \Pr(A)\Pr(B) \tag{3.6}$$

Definition

Two events A and B are said to be *conditionally independent* of an event E if, and only if,

$$\Pr(A \text{ and } B \mid E) = \Pr(A \mid E)\Pr(B \mid E) \tag{3.7}$$

3.2.4 *The law of total probability*

Finally, we use all three of these laws to define the law of total probability.

Theorem

The law of total probability: if C and $\bar{C}$ are mutually exhaustive events, then for some event *E*,

$$\Pr(E) = \Pr(E \mid C)\Pr(C) + \Pr(E \mid \bar{C})\Pr(\bar{C}) \quad (3.8)$$

3.2.5 Bayes Theorem

Theorem

If C and $\bar{C}$ are mutually exhaustive events, then for some event *E*,

$$\Pr(C \mid E) = \frac{\Pr(E \mid C)\Pr(C)}{\Pr(E \mid C)\Pr(C) + \Pr(E \mid \bar{C})\Pr(\bar{C})} \quad (3.9)$$

This definition is the traditional statistical definition, which does not lend itself easily to description. However, using the laws of probability, it is possible to restate Bayes Theorem using odds.

3.2.6 The relationship between probability and odds

We often hear about odds when we are talking about betting. In everyday speech, odds and probability tend to be used interchangeably; this is a bad practice because they are not the same at all.[102]

If we have some event C, then the odds in favor of C are given by

$$O(C) = \frac{\Pr(C)}{\Pr(\bar{C})} = \frac{\Pr(C)}{1 - \Pr(C)} \quad (3.10)$$

The odds on C are given by a ratio of two probabilities. This ratio can range from 0 (when Pr(C) = 0) to infinity (when Pr(C) = 1). When the odds are 1, they are referred to as *evens*. Let us examine our coin tossing example once again. What are the odds on observing a head from a single toss? If C is the event defined by "a head," then as we have seen Pr(C) = 0.5. So the odds on C are

$$O(C) = \frac{0.5}{1 - 0.5} = \frac{0.5}{0.5} = 1 \quad (3.11)$$

i.e., the odds are even. What about the odds on getting a card with a face value of 10 or higher on a single draw from a fair deck (52 cards, no jokers)? C = {10, J, Q, K, A}, so Pr(C) = 5/52, and the odds are

$$O(C) = \frac{\Pr(C)}{1 - \Pr(C)} = \frac{5/52}{47/52} = \frac{5}{47} \tag{3.12}$$

When the odds are less than one it is customary to invert them and call them *the odds against* C.[102] Here the odds are 47 to 5 against C. The concept of odds can be extended to conditional odds. That is, for some event C and some conditioning event E, the odds in favor of C given E are

$$O(C \mid E) = \frac{\Pr(C \mid E)}{\Pr(\bar{C} \mid E)} = \frac{\Pr(C \mid E)}{1 - \Pr(C \mid E)} \tag{3.13}$$

Once again, one can argue that all odds are conditional.

3.2.7 *The odds form of Bayes Theorem*

Theorem

If C and $\bar{C}$ are mutually exhaustive events, then for some event E,

$$\frac{\Pr(C \mid E)}{\Pr(\bar{C} \mid E)} = \frac{\Pr(E \mid C)}{\Pr(E \mid \bar{C})} \times \frac{\Pr(C)}{\Pr(\bar{C})} \tag{3.14}$$

In the following proof it can be seen that the condition that C and $\bar{C}$ be mutually exhaustive is not necessary for this form of Bayes Theorem. Several authors have put forward arguments justifying the relaxation of this condition. However, these arguments have little or no effect on the overall result.

Proof

From the third law of probability

$$\Pr(C \text{ and } E) = \Pr(C \mid E)\Pr(E) \text{ and } \Pr(\bar{C} \text{ and } E) = \Pr(\bar{C} \mid E)\Pr(E) \tag{3.15}$$

and, conversely,

$$\Pr(E \text{ and } C) = \Pr(E \mid C)\Pr(C) \text{ and } \Pr(E \text{ and } \bar{C}) = \Pr(E \mid \bar{C})\Pr(\bar{C}) \tag{3.16}$$

But,

$$\Pr(C \text{ and } E) = \Pr(E \text{ and } C) \text{ and } \Pr(\bar{C} \text{ and } E) = \Pr(E \text{ and } \bar{C}) \tag{3.17}$$

so

$$\Pr(C \mid E)\Pr(E) = \Pr(E \mid C)\Pr(C) \text{ and } \Pr(\bar{C} \text{ and } E)\Pr(E) = \Pr(E \mid \bar{C})\Pr(\bar{C}) \quad (3.18)$$

Now if we divide these two to find the odds on C given E, we get

$$\frac{\Pr(C \mid E)}{\Pr(\bar{C} \mid E)} = \frac{\Pr(E \mid C)}{\Pr(E \mid \bar{C})} \times \frac{\Pr(C)}{\Pr(\bar{C})} \quad (3.19)$$

The odds form of Bayes Theorem lends itself to a very simple method for updating prior belief. We may rewrite the expression 3.14 as

$$\text{Posterior odds} = \text{Likelihood ratio} \times \text{Prior} \quad (3.20)$$

As it has been demonstrated by various authors,[96,103] the evidence E alone does not allow the expert to give his/her opinion on C or $\bar{C}$. However, using Bayes Theorem it is possible to show how the evidence influences the probabilities associated with the two competing hypotheses.

Therefore, the likelihood ratio (LR) converts the prior odds in favor of C into the posterior odds in favor of C. It is this LR that the scientist will attempt to estimate in order to evaluate the value of evidence. It has to be noted that the background information I has been omitted for the sake of brevity, but it must be remembered that the probability of the event can only be measured in light of the circumstances of the case.* If the information I is explicitly included, then

$$\underbrace{\frac{\Pr(C \mid E, I)}{\Pr(\bar{C} \mid E, I)}}_{\text{posterior odds}} = \underbrace{\frac{\Pr(E \mid C, I)}{\Pr(E \mid \bar{C}, I)}}_{\text{LR}} \times \underbrace{\frac{\Pr(C \mid I)}{\Pr(\bar{C} \mid I)}}_{\text{prior odds}}. \quad (3.21)$$

Let us go back to the child abuse example. Let the event C be that the child has been abused and the evidence E be that the child rocks. We are told that 3% of abused children rock, i.e., that the probability that a child will rock given he or she has been abused is 3%, $\Pr(E \mid C) = 0.03$. If we are told that 3% of children who have suffered no abuse rock as well, $\Pr(E \mid \bar{C}) = 0.03$, then the odds form of Bayes Theorem tells us that

* One of the problems is that the suspect usually has no obligation to give any information or that the information given may not be admissible; therefore, the information I may be provisional.

$$\text{Posterior odds} = \frac{\Pr(E \mid C)}{\Pr(E \mid \bar{C})} \times \text{Prior odds}$$

$$= \frac{0.03}{0.03} \times \text{Prior odds} \quad (3.22)$$

$$= 1 \times \text{Prior odds}$$

$$\text{Posterior odds} = \text{Prior odds}$$

We now have indisputable, mathematically justified proof that the evidence does not give support to either hypothesis. This simple but startling revelation dispels all problems with coincidence thinking and also reveals how all evidence may be viewed in a legal context. All that remains for this chapter to show is how this philosophy may be applied.

3.3 *Bayesian thinking in forensic glass analysis*

In Evett's first article specifically on the use of Bayesian approach to glass evidence,[6] Bayes Theorem was applied at the comparison stage. However, as the method was not suitable for immediate application to casework, Evett and Buckleton[7] presented a pragmatic approach in order to benefit from the Bayesian philosophy while awaiting ideal treatments.

The authors made three compromises. First, they assumed that there existed a criterion in order to determine if the fragments did or did not match (for example, Student's *t*-test). Second, they assumed that the recovered particles had been grouped before treatment. Third, it was also assumed that there existed a means to estimate the frequency of occurrence of the observed characteristics.

As the last approach is less complex, we will present first the work that has been done on the use of Bayesian rationale after conventional tests and then proceed with the continuous approach presented in Evett,[6] Lindley,[5] or Aitken.[96] In order to illustrate the interaction of the various aspects of glass evidence, four different hypothetical cases will be presented: the first where one control and one group have been recovered and the other cases where multiple controls have been recovered.

Let us consider some of the thoughts that might occur to us *before* we start the examination of the case. At this early stage it is worth attempting to anticipate the opposing views that might be presented if the case did come to court. The prosecution will present a simple view.

The suspect committed the offense.

The defense is not compelled to offer an alternative, but it may be reasonable to suppose a sensible alternative, at least *a priori*, might be

The suspect did not commit the offense.

It is necessary to approach our examinations with these opposing views in mind. In doing so, we recognize that they are provisional and might change with changing circumstances. We see here the definition of two opposing hypotheses, a requirement for the Bayesian approach, and we suggest one of the fundamental principles of evidence interpretation: evidence may only be interpreted in the context of two (or more*) hypotheses. In 1997, Cook et al. (personal communication, 1997) discussed the hierarchy of propositions that can be addressed in a case. These propositions fall into three broad generic classes which are called source, activity, and offense level. The offense level is generally the level the court addresses. The two views of the defense and prosecution hypotheses at the offense level are given in the previous example. It is customary for forensic scientists to avoid such issues. Indeed, to address the offense level there is usually more than one piece of evidence, and the judge or jury is more competent to weigh both propositions. Hypotheses tested on the activity level can be addressed by the criminalist. The propositions could be as follows:

C: the suspect smashed the window.
$\bar{C}$: the suspect did not smash the window.

This level of the hierarchy presupposes that parameters such as transfer and persistence can be estimated. If they cannot, then only the source level can be addressed. The propositions of this last level are as follows:

C: the glass comes from the broken window.
$\bar{C}$: the glass comes from another broken glass object.

When addressing the lowest level, it must be stressed that the value of evidence will be either over- or underestimated. Therefore, in our opinion, if nothing is known on the circumstances of the case, the expert should then seriously consider not working the case.

Let us suppose that we address the activity level. We should, at this stage, be asking ourselves what help we might be able to give in resolving the conflict between these opposing hypotheses. The circumstances tell us that if C is the case, then we can expect to find glass on the suspect's clothing which will match the control sample. How much glass?

- No glass?
- One or two small fragments?
- A few small fragments?
- Lots of small fragments?

* As there may be more than two hypotheses, Robertson and Vignaux[94] suggested using $H_1, H_2, \ldots, H_n$ instead of C and $\bar{C}$.

- One or two big fragments?
- A few big fragments?
- Lots of big fragments?
- A mixture of small and big fragments?
- None of these?

What if $\bar{C}$ is the case? Bear in mind what we have been told about the suspect.

Given the circumstances, what strength of evidence is likely to be reported if C is actually the true hypothesis? If our expectation is that we are only likely to find one or two small fragments, then we may not be able to give more than a cautious opinion. If so, does the case examination represent good value for money (Cook et al., personal communication, 1997).

What results are we likely to arrive at if $\bar{C}$ is actually the true hypothesis? What if we had been told that the suspect has previous convictions for this kind of offense? Whereas this would make him a good suspect from the police perspective, it may well confound things with respect to evidence transfer. If the suspect is a habitual law breaker, is it more or less likely that we will find various kinds of glass on his clothing? Is he at greater risk of being mistakenly associated with the scene than the average man in the street?

Later, we will consider probabilities of transfer and persistence within the context of this hypothetical case. However, the important point that we want to make here is that we should think about such probabilities as much *before* we find anything as after for two reasons: (1) because our expectations should determine what, if anything, we are going to examine and how and (2) because our post hoc assessments are likely to be colored by what we have actually found. If one says "yes, that is just what I would have expected," it is not nearly as convincing as being able to demonstrate that the findings were in accord with one's predictions.

Suppose search of the clothing revealed a number of fragments of glass on the surface. For the sake of generality we will not discuss, at this point, any particular number, but rather state that n (= 1,2,3,...) fragments were recovered from the suspect's clothing. Similarly, let us avoid, for the time being, the statistics of the measurements and take it that there was a clear match between them and the measurements on the control.

The Bayesian (compromise) approach[7] proceeds as follows.

Case 3.3.1 — One group, one control

Based on the eyewitness evidence and what is known about the suspect, we visualize the *prior odds* in favor of C:

$$\text{Prior odds} = \frac{P(C \mid I)}{P(\bar{C} \mid I)} \tag{3.23}$$

where I denotes all the nonscientific evidence which the court will take into account. Let E denote the event that we have found n fragments on the surface of the suspect's clothing which match the control. Then we are interested in the new posterior odds:

$$\text{Posterior odds} = \frac{P(C \mid E, I)}{P(\bar{C} \mid E, I)} \tag{3.24}$$

That is, the odds given that there is some evidence, E, is known in addition to I which may lend support to C or $\bar{C}$. This is where we use Bayes' Theorem, which tells us that

$$\frac{P(C \mid E, I)}{P(\bar{C} \mid E, I)} = \frac{P(E \mid C, I)}{P(E \mid \bar{C}, I)} \times \frac{P(C \mid I)}{P(\bar{C} \mid I)} \tag{3.25}$$

This expression identifies a crucial factor, the LR:

$$\text{LR} = \frac{P(E \mid C, I)}{P(E \mid \bar{C}, I)} \tag{3.26}$$

To evaluate the LR we must answer two questions:

What is the probability of the evidence given that the prosecution hypothesis is correct and given the background information?

What is the probability of the evidence given that the defense hypothesis is correct and given the background information?

These form the numerator and denominator of the LR. We will look at the denominator first. The question, in more detail, is

What is the probability that we would find a single group of n matching fragments on the surface of the suspect's clothing if he had not smashed the window at the scene, given what we know about the incident and the suspect?

One problem, of course, is that we have no information about the suspect's activities in the period leading up to the arrest, nor is he under any obligation to provide such details in many jurisdictions. Furthermore, we cannot be sure that *any* of the information given to us will be admissible; but we have to start somewhere, recognizing that our adopted I is provisional.

The way in which we proceed will be determined not only by the circumstances of the case, but also by the expertise, knowledge, and data that we have available to us. For the rest of this section we will base our analysis on the Lambert, Satterthwaite, and Harrison (LSH) clothing survey.[104] This survey will be discussed in more detail in Chapter 4. However, at this stage we need to know that this survey can be used to give estimates of nonmatching (random or background) glass on persons suspected of a crime. Regardless of what we know about the suspect, it is inarguable that he/she has come to police attention in connection with a breaking offense. The casework clothing survey is a survey of people who also have come to police notice in that way. This is a persuasive argument for considering this survey to be the most relevant in this case. Let us see how we might use the results. Let us break down E as follows:

1. One group of fragments was found on the surface of the suspect's clothing.
2. The group contained n fragments.
3. The group matched the control.

Using this breakdown, let us rephrase our question in relation to the denominator:

> *If we examine the clothing of a man who has come to police notice on suspicion of a breaking offense, yet he is unconnected with the offense, what is the probability that we will find a single group of n fragments which match the control in this particular case?*

The LSH survey distinguished between groups of glass that matched casework controls and those that did not. If we assume that the matching fragments in each of these cases did, in fact, come from the incident being investigated, then the nonmatching glass gives us a picture of the background glass that we can expect to see in such cases, when the suspect is unconnected with the crime being investigated.

- Let P_i denote the probability that an innocent police suspect will have i ($i = 0, 1, 2,...$) groups of glass on the surface of his clothing.
- Let S_j denote the probability that a group of glass fragments on the surface of such a person's clothing will consist of j fragments ($j = 1, 2,\ldots, n$)
- Let f denote the probability that a group of glass fragments on the surface of such a person's clothing will match in RI the control in this particular case.

To help in our analysis we make two further assumptions.

A4: there is no association between the number of groups of glass found on a person's clothing and the sizes of those groups.
A5: there is no association between the frequency of a given RI of glass on clothing with either the number of groups or the size of the group.

It is unlikely that either of these assumptions is exactly true; however, we believe they are at least as correct as any first order approximation. Experimental data suggest that there are no obvious strong correlations. We note, defensively, that it is not that this approach requires *more* assumptions than others, but rather it makes it easier for us to be clear about what the assumptions are.

By invoking A4 and A5 and using our new terminology, we are now able to answer our question in a logical fashion. That is, what is the probability that we will find a single group of n fragments which match the control sample in this particular case on a person unconnected with the crime?

$$P\left(E \mid \bar{C}, I\right) = P_1 . S_n . f \tag{3.27}$$

Using the new data collection, we are able to address a question in relation to the denominator, which is relevant to our original hypothesis, $\bar{C}$. This means that we can now turn our attention to the original C and address the numerator $P(E \mid C, I)$. The question, in more detail, is

> *What is the probability that we would find n matching fragments on the surface of the suspect's clothing if he had smashed the window at the scene, given what we know about the incident and the suspect?*

The numerator is complicated slightly by the need to allow for at least* two possible explanations for the evidence if C is the case:

1. *Either* the group of fragments was transferred from the scene window — in which case, the suspect could not have had any other glass on his clothing before.
2. *Or* no glass was transferred from the scene window, but the suspect already had the group of glass fragments on his clothing.

Let T_k denote the probability that, given C and I, k ($k = 0, 1, 2, \ldots, n$) glass fragments would be found from the scene window on the suspect's clothing. Then, again invoking A4 and A5,

* Actually, there are $n + 1$ explanations: r fragments were transferred, $n - r$ being there beforehand; $r = 0, 1, 2, \ldots, n$. But it is possible to show that most of the terms associated with these alternatives are small and leaving them out is, in any case, conservative, in the sense that we are making the numerator smaller than it should be.

$$P(E \mid C, I) = T_0.P_1.S_n.f + T_n.P_0 \tag{3.28}$$

The first combination of terms deals with the second of the two alternatives: no glass being transferred, T_0; one group of n fragments being there beforehand, $P_1.S_n$; which matched the control f. The second deals with the other alternative: n fragments were transferred and none were there beforehand; we assume, for the time being, that if glass fragments were transferred from the scene then they would match with probability 1. This is an approximation, but not a bad one.

We now have both the numerator and the denominator terms. Therefore, the LR is

$$\frac{P(E \mid C, I)}{P(E \mid \bar{C}, I)} = T_0 + \frac{P_0.T_n}{P_1.S_n.f} \tag{3.29}$$

Case 3.3.2 — Two recovered and one control groups

The crime, arrest, and exhibits are the same as in Case 3.3.1. The only difference is that ten fragments were found on the clothing surface. Based on RI, these formed two clear groups: one of six fragments that matched the control, and another of four fragments that was different from the control.

Again, our evaluation centers on the LR:

$$\text{LR} = \frac{P(E \mid C, I)}{P(E \mid \bar{C}, I)} \tag{3.30}$$

Once again, we will look at the denominator $P(E \mid \bar{C}, I)$ first and will use the LSH data. The question now is

> *If we examine the clothing of a man who has come to police notice on suspicion of a breaking offense, yet he is unconnected with the offense, what is the probability that we will find two groups of fragments of the observed sizes and properties?*

Let f_1 and f_2 denote the probabilities that groups of glass found on clothing will have the RIs of the two observed groups. Let group 1 be the matching group and group 2 be the nonmatching group. Then,

- P_2 is the probability that the suspect's clothing would have two groups of glass.
- $S_6 f_1$ is the probability that a group of glass fragments on clothing will have six fragments and the observed RI of group 1.

- S_4f_2 is the probability that a group of glass will have four fragments and the observed RI of group 2.

Again, we are using assumptions A4 and A5, leading to

$$P(E|\bar{C},I) = 2P_2S_6S_4f_1f_2 \tag{3.31}$$

The factor 2 arises because if we sample two groups of glass from clothing there are two ways in which we can meet our requirements: the first group can be size 6 and the second group size 4, or the first group can be size 4 and the second group size 6.

As in the first case, when we consider the numerator we have to allow for two possibilities:

1. *Either* no glass was transferred from the scene window, but the suspect already had the two groups of glass fragments on his clothing.
2. *Or* the group of six fragments was transferred from the scene window — in which case, the suspect had a single group of four fragments on his clothing beforehand.

Then,

$$P(E|C,I) = T_0.2P_2.S_6.S_4.f_1.f_2 + T_6.P_1.S_4.f_2 \tag{3.32}$$

So the LR is

$$\text{LR} = \frac{P(E|C,I)}{P(E|\bar{C},I)} = T_0 + \frac{P_1.T_6}{2P_2.S_6.f_1} \tag{3.33}$$

Unlike the conventional approach using the coincidence concept,[92] the frequency of the nonmatching group does not appear in the final evaluation. Moreover, the value of the evidence is different if the groups are constituted of six matching and four nonmatching fragments or four matching and six nonmatching fragments. One can also see that the size of the nonmatching group is irrelevant.

Case 3.3.3 — One recovered and two control groups

The scenario is the same as before, but two windows have been broken and only one group of glass has been recovered on the suspect's clothing. This recovered glass matches control group 1. In order to estimate the probability of coincidence, Evett and Lambert[92] suggested summing the coincidence probabilities estimated for each control, so the value of the evidence remained the same whether the recovered glass matched the more or the less frequent glass. This problem does not arise in the Bayesian framework.

If the suspect was present when the window was broken, there are two possibilities.

1. Sample 1 was transferred and recovered, but not Sample 2.
2. The glass is present at random, neither Sample 1 or 2 has been transferred from the scene windows.

If the suspect was not present when the window broke, then the glass is present at random.

Assuming that

- T_{1k}: The probability that k fragments have been transferred from control 1 have persisted and have been recovered.
- T_{2l}: The probability that l fragments have been transferred from control 2 have persisted and have been recovered.
- P_i: The probability of finding i groups at random on the suspect's clothing
- $S_n f_1$: The probability that a group of glass fragments on clothing will have n fragments and the observed characteristics of group 1.

Then the LR is

$$\text{LR} = \frac{P(E \mid C, I)}{P(E \mid \bar{C}, I)} = \frac{T_{1_0} T_{2_0} P_1 S_n f_1 + T_{1_k} T_{2_0} P_0}{P_1 S_n f_1} = T_{1_0} T_{2_0} + \frac{T_{1_k} T_{2_0} P_0}{P_1 S_n f_1} \tag{3.34}$$

If the probability of transfer is the same for controls 1 and 2, then

$$\text{LR} = T_0^2 + \frac{T_k T_0 P_0}{P_1 S_n f_1} \tag{3.35}$$

Intuitively one would think that control 2 has to be taken into account; indeed, the more controls the more probable it seems to have a match due to chance. However, it is not the frequency of the intrinsic characteristics (RI or other measurements on control 2) that influences the value of evidence, but the fact that no glass has been recovered! It is T_{20} (in this case T_0) that diminishes the LR. Note also that unlike the coincidence approach,[92] the LR changes with the frequency of the intrinsic characteristics of the matching group; if the recovered glass is rare, the LR is higher than if the recovered glass is common.

Case 3.3.4 — Two recovered and two control groups

The crime, arrest and circumstances are the same as in the case presented before, but two groups of glass fragments have been recovered on the suspect's garments. These two groups match the two controls. If we consider the denominator, the two groups are present at random. For the numerator, we have to allow for four possibilities.

1. The two groups of glass were transferred from scene windows one and two, and the suspect had no glass on his clothing beforehand.
2. One group of glass came from scene window 1, no glass was transferred from scene window 2, but the suspect already had one group of glass on his clothing.
3. One group of glass came from scene window 2, no glass was transferred from scene window 1, but the suspect already had one group of glass on his clothing.
4. No glass was transferred from the two scene windows, but the suspect already had the two groups of glass on his clothing beforehand.

So the LR is

$$\mathrm{LR} = \frac{P(E \mid C, I)}{P(E \mid \bar{C}, I)} = \frac{T_{1_k} T_{2_l} P_0 + T_{1_0} T_{2_l} P_1 S_n f_1 + T_{1_k} T_{2_0} P_1 S_m f_2 + T_{1_0} T_{2_0} . 2 P_2 S_n S_m f_1 f_2}{2 P_2 S_n S_m f_1 f_2}$$

$$\mathrm{LR} = \frac{T_{1_k} T_{2_k} P_0}{2 P_2 S_n S_m f_1 f_2} + \frac{T_{1_0} T_{2_l} P_1}{2 P_2 S_m f_2} + \frac{T_{1_k} T_{2_0} P_1}{2 P_2 S_n f_1} + T_{1_0} T_{2_0} \qquad (3.36)$$

If the probability of transfer is the same for controls 1 and 2, then

$$\mathrm{LR} = \frac{T_k^2 P_0}{2 P_2 S_n^2 f_1 f_2} + \frac{T_0 T_k P_1}{2 P_2 S_n f_2} + \frac{T_k T_0 P_1}{2 P_2 S_n f_1} + T_0^2 \qquad (3.37)$$

where $S_m f_2$ is the probability that a group of glass fragments on clothing will have n fragments and the observed characteristics of group 1.

The Bayesian approach, presented earlier in the different cases, shows which parameters are important to consider when evaluating the value of a match. It allows us also to explain what is the correct population to consider, for example, why one should take into consideration the frequency of intrinsic characteristics of glass found at random and not of control glass. The different scenarios demonstrate that even if there are several controls and recovered glass and/or when there are nonmatching groups, the Bayesian framework gives logical answers unlike the conventional approach.

3.3.1 A generalized Bayesian formula

One of these terms typically predominates in cases with matching glass. Therefore, it is possible to give a simple approximate general formula:

$$\mathrm{LR} = \frac{(G-M)!\,P_{G-M}T_{s_1}T_{s_2}\ldots T_{s_m}T_0^N}{G!\,P_G S_{s_1}S_{s_2}\ldots S_{s_m}f_1 f_2\ldots f_m} \tag{3.38}$$

where

- G is the total number of groups of glass on the clothing.
- M is the number of matching groups.
- N is the number of controls not apparently transferred. (A different T_0 is realistically expected for each control if the circumstances of each breakage differ in any way. If such an extension is desired, the term T_0^N should be replaced by the product of N T_0 terms.)
- T_{si} is the transfer probability for the group i of size S_i.
- S_i is the probability of a group of size s_i.
- f_i is the frequency of the ith group.

This formula may be easily programmed into Microsoft Excel.

3.4 Taking account of further analyses

So far, the data we have considered consists solely of RI measurements. In practice, we can often gather more data on both recovered and control fragments. The methods of annealing and elemental analysis are widely used in this regard. How do we take account of this additional data? Essentially, there are two ways in which the added information can change our assessment. The first is that we may gain added discrimination from this information (assuming that a match results).

Recall that we defined f as "the probability that a group of glass fragments on the surface of a suspect's clothing will match in RI in this particular case." We could, for example, redefine f as "the probability that a group of glass fragments on the surface of a suspect's clothing will match in RI and *elemental analysis* in this particular case." Assuming that elemental analysis does indeed result in added discrimination, then the new f will be much smaller than the old f and the LR will be bigger. However, in order to estimate f, we must have a clothing survey that includes the results of elemental analysis, requiring equipment which, at present, most laboratories do not possess.

The second way in which additional information can affect our conclusions is potentially much more complicated. The additional data can affect the hypotheses we consider. In particular, the new data can affect the way in which we will react to future provisional hypotheses that may be put forward in court. The obvious example is where defense presents, possibly as mere speculation, an alternative explanation for the presence of matching glass. The most obvious explanation is that the defendant had recently been in the vicinity of some other breaking glass object. If elemental analysis enables us to classify the recovered fragments, then the probability of the evidence under alternative hypotheses may be reduced or enhanced.

Imagine that the suspect says that he broke a beer jug recently. If either interferometry or elemental analysis demonstrates that one or more of the recovered fragments are from a flat (or flat float) source, then the probability of this evidence, if the control glass is from a beer jug, is reduced (to zero if we accept that the results are always correct and that all the recovered fragments have come from one source). What if the elemental or interferometry analysis suggests that at least one fragment may have come from a container source? It may be tempting to consider this fragment "eliminated" and to proceed with assessing the remaining fragments in some way that supports the prosecution hypothesis. We cannot support this approach. The presence of even one confirmed container fragment in the group at this RI greatly increases the support for the alternative hypothesis. We will develop these hypotheses at more length later.

3.5 Search strategy

Extending Bayesian arguments suggests a logical method for search strategies and stopping rules. Consider the simplest formula given here: for one control and one matching recovered group

$$\frac{P(E|C,I)}{P(E|\bar{C},I)} = T_0 + \frac{P_0 \cdot T_n}{P_1 \cdot S_n \cdot f} \tag{3.39}$$

This formula is large whenever the ratio $\frac{P_0}{P_1 \cdot S_n}$ is large. The terms in this formula relate to

- P_0: the probability of finding no glass on the clothing of a random person following THIS search strategy.
- $P_1 S_n$: the probability of finding one group of glass of size n on the clothing of a random person following THIS search strategy.

Therefore, this ratio is most likely to be maximized by the following strategy. The order of the examination should be

1. Hair combings
2. The surface of clothing
3. The surface (the uppers) of shoes
4. Pockets and cuffs of clothing
5. Glass embedded in shoes

This order is derived from an examination of Table 2 in both Harrison et al.[105] and Lambert et al.[104]

Searching should be stopped at some point. This point should be when "most" of the evidence has been established. Chapter 4 will show that groups of glass of size 3 or over are rare on persons unconnected to a crime. This suggests that searching should be stopped when more than three fragments have been found, and stopping when ten or more pieces of glass in total have been recovered is certainly justified, especially if the next search would be to move to a lower level in the hierarchy given previously.

It could be argued with some justification that this policy is designed to optimize evidence from the prosecution point of view and, therefore, is biased against the defendant. However, this same search strategy avoids those areas most likely to produce "random" glass and, hence, to some extent safeguards an innocent suspect against the largest risks of coincidental matching.

No search strategy can safeguard an innocent suspect who coincidentally has large amounts of matching glass on his/her hair or upper clothing.

This policy suggests that if, say, 20 fragments are found on the T-shirt, the first garment searched, then about 10 of these should be examined. At this point the control may be opened and the number of fragments recommended in Chapter 2 should be examined from the control (ten control fragments if there are ten measured recovered fragments). If these "match," then the examination should be stopped. We seriously recommend stopping without opening the lower clothing or shoes in such a case. This optimizes evidence and saves time.

In the event that a high probability of transfer is expected and no glass is found on the T-shirt, we also recommend stopping and writing a statement supporting the suggestion that the clothing was NOT close to the window when it was broken. The control should be examined after the search in case there is evidence (such as the control is wired or laminated) that suggests that much effort would have to be exerted to penetrate it, thereby increasing the belief in high transfer probabilities.

We suggest that only in special circumstances should a search continue all the way to the soles of the shoes (however, this matter should also be discussed with the judge or police). The only circumstance that comes to mind is that the window was broken by throwing an object from a distance and the only suspected contact was the offender walking (gently) over the glass.

3.6 Comparison of measurements: the continuous approach

In this section we introduce a method known as the continuous approach. Under match/nonmatch thinking (as with Student's *t*-test), the evidence is a match until some predetermined point where it suddenly becomes a non-match. Smalldon (personal communication, 1995) termed this "the fall off the cliff effect." Under this thinking, a "good" match is assessed as having the same evidential value as a "poor" match. A very narrow mismatch is reported in the same way as an obvious mismatch. It is possible that the glass did not pass a reasonable statistical test, but the RIs are still quite close and there is the fact that there may be a lot of glass on the clothing.

The faults lie in the match/nonmatch approach and the sequential way in which evidence is often considered. The solution lies in abandoning match thinking, as suggested by Lindley.[5]

Abandoning the match/nonmatch approach is one of the great advances of inferential thinking in forensic science. It has several advantages. Philosophically, it means that the forensic scientist is permitted to weigh all the evidence in one process rather than in some step-by-step fashion.

We perform here a thought experiment. Imagine some comparison process between a control and recovered sample of something. Scientist 1 performs test 1 first. At test 1 the two samples narrowly mismatch using some rule that has an error rate α. Most scientists are obliged to stop here and report a mismatch, and indeed scientist 1 does so.

Scientist 2 starts at test 2 for some reason and finds a correlation of features that are otherwise very rare in the population. She then performs test 3 and finds another correlation of features that is otherwise rare. Tests 4, 5, and 6 also find a correlation of features that is otherwise very rare in the population. Scientist 2 then finally performs test 1 and finds the narrow mismatch. Such a scientist is presented typically with a very difficult problem. Most of the evidence suggests an association; however, one test has found a mismatch. This test, however, is known to have a small error rate. Has an error occured here? The match/mismatch approach is unable to recover from this situation. In fact, the more tests that are performed on truly matching samples, the more likely such an ambiguity will result by a false exclusion occurring in one of the tests. How can this be? Surely more testing must be better? The fault is in the match/nonmatch approach. This approach may also be compared to the logic of Karl Popper. In this logical system a hypothesis is advanced and subsequent tests attempt to disprove it. The more tests the hypothesis survives, the more credible the hypothesis. The forensic corollary of this hypothesis is that the two samples have the same source. This is the logical system many of us were taught. However, return to scientist 2. We suspect that her true belief is that the evidence supports an association and that the close mismatch is an error in this case. She may be obliged to report a mismatch or she may (indeed, in single locus work in

DNA in the U.S. she might) report the close mismatch as inconclusive and proceed to interpret the remaining tests. Neither answer is correct. The correct approach lies, as we have suggested, in abandoning the match/non-match approach.

How can this be done? The method is no surprise to professional statisticians, but typically seems completely unfamiliar to forensic scientists. Therefore, we spend some time introducing it.

Imagine another thought experiment. Scientist 3 has the same problem, except that she is equipped with the scales of justice and a blindfold. Every time she interprets a test, she puts the correct weight on the scales either on the defense side or the prosecution side as is warranted. For test 1, the narrow mismatch, she puts the correct weight on the defense side. For tests 2 to 6 she puts the weights on the prosecution side. In the end she simply reports the resulting net weight and which side it supports.

If we have convinced anyone that this latter approach is highly desirable, then all that is left is to show how to implement it. The implementation of this approach is standard statistical methodology.

The basis of the implementation is the use of probability density rather than probability itself. The concept of a probability density is novel to many forensic scientists and seems very difficult to explain. One reason for this is that it is often not taught at school, and indeed when it is taught, typically, it is mistaught.

Most of us have been shown a normal distribution at school, possibly something like Figure 3.1. This is intended to be a distribution of men's heights. It peaks somewhere just below 6 ft. The next question is simple since we are all so familiar with this graph. What is the y-axis?

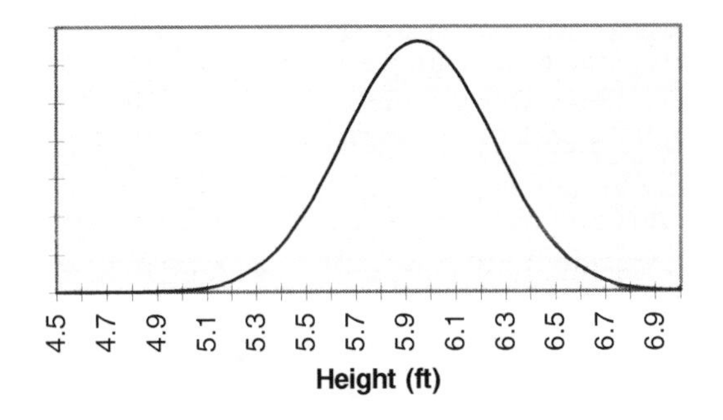

Figure 3.1 A distribution of men's heights.

We cannot tell what you answered or whether you realized that you have probably never been told or shown this. If you answered probability then you were wrong. The areas under this graph are the probabilities. Perhaps you answered counts or relative frequency, both of these are also

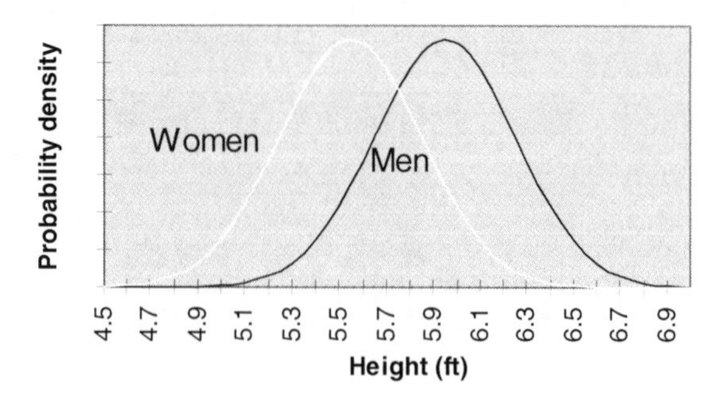

Figure 3.2 A distribution of men's and women's heights.

wrong (although relative frequency might be close). The *y*-axis is probability density. This tells us that there is more probability about the mean and less in the tails. It is not a concept that comes easily, but it is relatively easy to use. Another graph is shown in Figure 3.2.

The white line is intended to be a survey of women's height superimposed on the previous one of men's heights. We can ask various questions about this graph, some of which are easy to answer and some of which are not. At the risk of spreading confusion, we initially ask "what is the probability that a man is 6.3 ft?" This question is unanswerable. All that can be answered is how many men are between 6.25 and 6.35 ft (or a similar question). We must define a width to get an area on this graph. Once we define the width, we can answer the question. In this case the answer is 0.067444, which we have obtained by taking the area under the men's graph between 6.25 and 6.35 ft. Has this got anything to do with forensic science? What is the probability that a set of recovered fragments will have an RI of 1.5168? First, you must tell me the width you want (the match window perhaps?) then I can answer the question.

The need to define a width to calculate a probability leads inevitably to the need to define a match window and to troubles associated with that.

As another thought experiment let us try the following. I have a person who is 6.3 ft tall. Do you think it is a man or a woman? The evidence supports it being a man. Why? Because the men's graph is higher at the 6.3 ft mark than the women's. In fact, the exact support for this suggestion is the ratio of the heights of the two graphs at this point. The ratio in this case is 11.3. Can this concept be used in any way? It has the potential to avoid all questions about matching and match windows. It has the potential to weigh all evidence in the way suggested in the scientist 3 thought experiment. It has the potential to weigh close matches, poorer matches, and mismatches. This is done by constructing probability density distributions for glass evidence.

As a practical step in the implementation we make some simplifications. We propose to retain the grouping assumption for the benefits of simplicity that follow from it, but to drop the match/nonmatch approach. The future may involve dropping the grouping approach as well, but at this point we cannot implement that suggestion.

As a simple way of imagining this transition, recall the formula for the example given earlier for n matching fragments and no mismatching glass.

$$\frac{P(E \mid C, I)}{P(E \mid \bar{C}, I)} = T_0 + \frac{P_0 \cdot T_n}{P_1 \cdot S_n \cdot f} \tag{3.40}$$

This, we have previously suggested, can be approximated by

$$\frac{P(E \mid C, I)}{P(E \mid \bar{C}, I)} = \frac{P_0 \cdot T_n}{P_1 \cdot S_n \cdot f} \tag{3.41}$$

There is a $1/f$ term in this equation. This is the probability of a match if they are from the same source (1) divided by the probability of a match if they are not (f). It is this $1/f$ term that can be replaced by the respective probability densities. The numerator (previously 1) becomes a term relating to the closeness of the match and the denominator (previously f) becomes a term relating to the rareness of the glass. It can be seen that there is a very clear correspondence between the two approaches.

The formula in this case (to replace the $1/f$ term) is

$$\frac{P(\bar{x} - \bar{y} \mid \bar{x}, S_x, S_y, C)}{P(\bar{y} \mid \bar{x}, S_x, S_y, \bar{C})} \tag{3.42}$$

which can be interpreted as a numerator expressing the probability density that the two means of the recovered and control are this close, over a denominator of the probability density that the recovered mean is as observed.

In Appendix A, we give our justification for this formula. It will be noted that we have omitted for simplicity small positive terms such as T_0. The omission of these small positive terms has a negligible effect in most cases, but in any case this effect is to the advantage of the defendant.

We return to Case 3.1.1 which we reinterpret using the continuous approach. An excerpt of the probability density function of the RI of glass on clothing from the LSH survey is given in Figure 3.3. The probability density function of the difference of the sample means on the same scale is also given in Figure 3.3.

This distribution is used in Welch's modification of Student's t-test and assumes that the variances of the populations from which the two samples

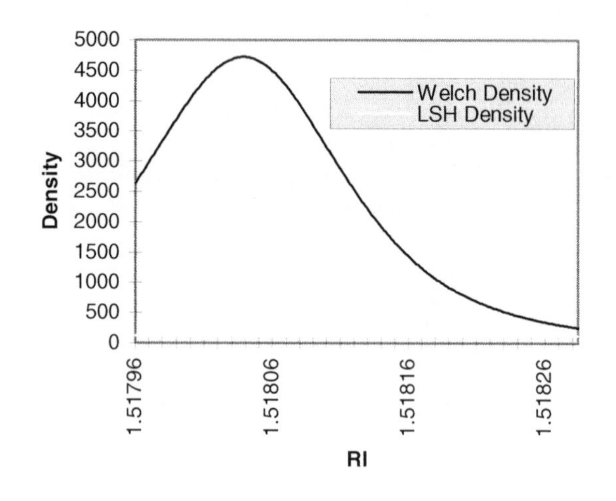

Figure 3.3 Welch and LSH densities in the area about RI = 1.5180.

are drawn are unknown and unequal. The justification for this appears in the appendix.* At the recovered mean 1.51804, the values are 4713 for the Welch density and 120 for the LSH density (see Figure 3.3). It can be seen that the LSH density is essentially a low horizontal line on this scale.

We now define:

where

- $f(\bar{y})$ is the value of the probability density for glass at the mean of the recovered sample (from Figure 3.3).
- $p(\bar{x} - \bar{y} \mid s_x s_y)$ is the value of the probability density for the difference between the sample means (from Figure 3.3).

As before, we take approximate values for the P and S terms from LSH

$$\mathrm{LR} = 0.079 + \frac{0.25 \times 0.042 \times 4713}{0.22 \times 0.02 \times 120} \qquad (3.43)$$

$$= 94$$

which compares with the value 60 previously obtained and reflects the close nature of the match.

* Strictly we wish to model the term $p(\mathbf{X}, \mathbf{Y} \mid C)$. The vectors **X** and **Y** are the observations from the control and recovered groups. This is, however, very complex, and solutions to date involve a large number of assumptions. Modeling the more assessable term given is not likely to differ in a meaningful way.

3.6.1 A continuous LR approach to the interpretation of elemental composition measurements from forensic glass evidence

In Chapter 2, Hotelling's T^2 test, a multivariate equivalent of Student's t-test, for determining a match between glass fragments recovered from a suspect and a control sample of glass fragments was introduced. While Hotelling's T^2 test is certainly a better approach than the 3 sigma rule, it is still subject to the weaknesses inherent in any hypothesis testing approach. Hypothesis testing suffers from three main problems in the forensic arena. The first problem is that hypothesis tests fail to incorporate relevant evidence, such as the relative frequency of the recovered glass and the mere presence of glass fragments. The second problem is what Smalldon (personal communication, 1995) termed the "fall off the cliff" effect. It seems illogical that if a probability of 0.989999 is returned from a hypothesis test then it should be deemed as a match when a probability of 0.990001 would be a nonmatch, particularly when these probabilities are calculated under distributional assumptions, which (almost certainly) do not hold. The third problem is that hypothesis testing does not answer the question of interest to the court. Robertson and Vignaux[106] argue that presentation of a probability answers the predata question — "What is the probability of a match if I carry out this procedure?" — rather than the postdata question — "How much does this evidence increase the likelihood that the suspect is guilty?" It is, of course, the latter that the court is interested in.

The suggestion to take a Bayesian approach to these problems is by no means novel,[5] and the continuous extension has been used in dealing with RI-based data.[8] The next section will extend the continuous approach for multivariate (elemental composition) data.

3.6.1.1 The continuous likelihood ratio for elemental observations

Walsh et al.[8] discuss a case where a pharmacy window was broken. Fragments of glass were retrieved from two suspects and compared with a sample of fragments from the crime scene on the basis of mean RI. Both recovered samples failed their respective t-tests and that is where the matter might have ended. However, there were a number of aspects contradictory to the conclusion that the fragments did not come from the crime scene window. Both offenders had a large number of fragments of glass on their clothing. Studies[104,107,108] have shown that large groups of glass fragments on clothing is a reasonably rare event on people unassociated with a crime. Examination suggested that the recovered fragments came from a flat float glass object — again a relatively rare event — and paint flakes recovered from one of the suspects were unable to be distinguished from the paint in the window frame at the crime scene.

Thus, the weight of the evidence supports the suggestion that both suspects were at the crime scene when the window was broken. This is a classic example of Lindley's paradox,[109] although Lindley himself refers to it as Jeffreys' paradox. Although the samples failed the t-test, the results are still more likely if they had come from the same source than from different sources.

Walsh et al.[8] propose an extension to the ideas put forward by Evett and Buckleton[7] which retains grouping information while dropping the match/nonmatch approach. The formula under consideration in their specific case is

$$\mathrm{LR} \approx \mathrm{T}_0 + \frac{T_L P_0 f(\bar{X} - \bar{Y} \mid S_X, S_Y)}{P_1 S_L \hat{g}(\bar{Y})} \tag{3.44}$$

where

- T_L = the probability of three or more glass fragments being transferred
- P_0 = the probability of a person's having no glass on their clothing
- P_1 = the probability of a person's having one group of glass on their clothing
- S_L = the probability that a group of glass on clothing contains three or more fragments
- $\hat{g}(\bar{Y})$ = the value of the probability density for float glass at the mean of the recovered sample, usually obtained from a density estimate
- $f(\bar{X} - \bar{Y} \mid S_X, S_Y)$ = the value of the probability density for the difference of two sample means; this is simply an unscaled t-distribution using Welch's modification to Student's t-test

The term T_0 is generally small and, hence, can be dropped without significant loss in generality or accuracy. Therefore, Equation 3.1 can be rewritten as

$$\mathrm{LR} \approx \mathrm{T}_0 + \frac{T_L P_0}{P_1 S_L} . lr_{cont} \tag{3.45}$$

where

$$lr_{cont} \approx \frac{f(\bar{x} - \bar{y} \mid S_x, S_y)}{\hat{g}(\bar{y})} \tag{3.46}$$

In a case where the glass evidence is quantified by elemental decomposition rather than by RI, the only change in evaluating the LR is the method for evaluating lr_{cont}.

Hotelling's T^2 is a multivariate analog of the *t*-test that examines the standardized squared distance between two points in *p*-dimensional space. These two points, of course, are given by the estimated mean concentration of the discriminating elements in both samples. It seems logical that the multivariate form of lr_{cont} should replace $f(\bar{X} - \bar{Y} \mid S_X, S_Y)$ with the unscaled probability density function for the distribution of T^2. This, however, is not quite as simple as it sounds. If there are n_c control fragments and n_r recovered fragments to be compared on the concentration of p different elements, and $n_c + n_r > p + 1$, then it is assumed that (1) the elemental data of the whole of the control population (window/container/bottle) and the whole of the recovered population is multivariate normal and (2) both populations have the same covariance structure,[89] i.e., the individual variances of the elements are the same in each population, and the correlation between any two elements is the same in each population. If these assumptions are true, then T^2 has an *F*-distribution scaled by the sample sizes, i.e.,

$$T^2 \sim \frac{(n_c + n_r - 2)p}{(n_c + n_r - p - 1)} F_{p, n_c + n_r - p - 1} = c^2 F_{p, n_c + n_r - p - 1} \tag{3.47}$$

where

$$c^2 = \frac{(n_c + n_r - 2)p}{(n_c + n_r - p - 1)} \tag{3.48}$$

Thus, $f(\bar{X} - \bar{Y} \mid S_X, S_Y)$ should be replaced by the value of the unscaled probability density for an *F*-distribution on p and $n_c + n_r - p - 1$ degrees of freedom at T^2/c^2, and $\hat{g}(\bar{Y})$ should be replaced by the value of a multivariate probability density estimate at the recovered mean $\bar{Y}$ (recall that $\bar{Y}$ is now a $p \times 1$ vector). However, the scaling factor in the multivariate case is a matrix, while both $f(T^2/c^2)$ and $\hat{g}(\bar{Y})$ are scalars, so lr_{cont} could be evaluated, but the result would be a matrix and have no intuitive meaning. The solution to this problem comes from the way the Hotelling's T^2 test works.

Hotelling's T^2 finds the linear combination of the variables that maximizes the squared standardized distance between the two mean vectors. More specifically, there is some vector l ($p \times 1$) of coefficients such that $l^T(\bar{X} - \bar{Y})$ quantifies the maximum population difference. That is, if T^2 rejects the null hypothesis of no difference, then $l^T(\bar{X} - \bar{Y})$ will have a nonzero mean.[88] This fact provides the solution. If l can be found (and it can, see Appendix A), then it can be shown that a new statistic t_l^2 (see Appendix A for the definition) has the same distribution as T^2,[88] but has a scaling factor that is not a matrix. The numerator of lr_{cont}, therefore, becomes $f(l^T(\bar{X} - \bar{Y}) \mid s_l)$. The numerator is the height of a probability density for an *F*-distribution on p and $n_c + n_r - p - 1$ degrees of freedom at t_l^2/c^2 transformed back to the scale of the linear combination. The denominator must be on the same scale, thus it becomes the value of a univariate probability estimate density at $l^T\bar{Y}$.

3.6.1.2 Examples

The data in the following examples come from two distinct sources, one green bottle and one colorless bottle taken from the same plant at the same time. Ten fragments were taken from each bottle, and the concentrations of aluminum, calcium, barium, iron, and magnesium ($p = 5$) were determined by ICP-AES. The quantities, T_L, P_0, P_1, and S_L, are taken to be those given in Reference 8, so that

$$\text{LR} \approx 8lr_{cont} \tag{3.49}$$

The first example uses five fragments from the green bottle as a control sample ($n_c = 5$) and five fragments from the same bottle as a recovered sample ($n_r = 5$) so that the population means are truly equal. $lr_{cont} \approx 2600$, so in this case the evidence would be 20,800 times more likely if the suspect was at the crime scene than if he was not.

The second example takes the ten fragments from the green bottle as the control sample ($n_c = 10$) and the ten fragments from the colorless bottle as the recovered sample ($n_r = 10$), so the null hypothesis is false, i.e., the population means are truly different. $lr_{cont} \approx 4 \times 10^{-10}$. Because the lr_{cont} is so small, the term T_0 in Equation 3.45 now determines the LR. If T_0 is taken at a typical value of around 0.1, then in this case the evidence would be ten times less likely if the suspect was at the crime scene than if he was not, i.e., the evidence against the suspect strongly disputes the hypothesis that the suspect was at the crime scene.

3.6.1.3 Discussion

Hotelling's T^2 test for the difference in two mean vectors provides a valid statistical method for the discrimination between two samples of glass based on elemental data. However, it is subject to small problems and does not answer the real question adequately. The Bayesian approach along with the continuous extension is the only method that fulfills the requirements of the forensic scientist, the statistician, and the court. All analyses of elemental data should use the continuous Bayesian approach.

3.7 Summary

In this chapter we have presented the laws of probability and introduced the Bayesian approach as the best available model for forensic inference. We have presented a hierarchy of propositions, which could be addressed by the court and/or expert. In order to illustrate the interaction of the various parameters that are important in assessing glass evidence, we have described four hypothetical cases. For these cases a compromise approach has been adopted, and because of the complexity of the continuous LR it is only at the end of the chapter that we have presented a full Bayesian approach.

The framework presented has many advantages, one of which is to highlight the sorts of surveys and experimental data that are needed to assess the value of glass evidence. Most of the parameters appearing in the LR are intuitively important to glass examiners, and some had been studied before it was shown that Bayesian inference provided a logical and coherent framework for transfer evidence. However, the great majority of these studies, particularly the surveys done on glass found on clothing at random, are posterior. These surveys will be the subject of the following chapter.

3.8 Appendix A

We require

$$\frac{p\left(\bar{x},\bar{y},s_x,s_y \mid C\right)}{p\left(\bar{x},\bar{y},s_x,s_y \mid \bar{C}\right)} = \frac{p\left(\bar{x}-\bar{y},\bar{x},s_x,s_y \mid C\right)}{p\left(\bar{x},\bar{y},s_x,s_y \mid \bar{C}\right)}$$

$$= \frac{p\left(\bar{x}-\bar{y} \mid \bar{x},s_x,s_y,C\right)}{p\left(\bar{y} \mid \bar{x},s_x,s_y,\bar{C}\right)} \times \frac{p\left(\bar{x} \mid s_x,s_y,C\right)}{p\left(\bar{x} \mid s_x,s_y,\bar{C}\right)} \times \frac{p\left(s_x \mid s_y,C\right)}{p\left(s_x \mid s_y,\bar{C}\right)} \times \frac{p\left(s_y \mid C\right)}{p\left(s_y \mid \bar{C}\right)}$$

$$= \frac{p\left(\bar{x}-\bar{y} \mid \bar{x},s_x,s_y,C\right)}{p\left(\bar{y} \mid \bar{x},s_x,s_y,\bar{C}\right)} \tag{3.50}$$

By assuming $p(\bar{x} - \bar{y} \mid \bar{x}, s_x, s_y, C) \cong p(\bar{x} - \bar{y} \mid s_x, s_y, C)$, we assume that the sampling variation for fragments with mean μ is independent of μ:

$$p\left(\bar{y} \mid \bar{x},s_x,s_y,\bar{C}\right) \cong p\left(\bar{y} \mid \bar{C}\right) \text{ and } \frac{p\left(s_x \mid s_y,C\right)}{p\left(s_x \mid s_y,\bar{C}\right)} \cong 1 \tag{3.51}$$

While this assumption is unlikely to be exactly true, the expected value of this ratio is expected to be greater than one. Replacing it by one is unlikely to disadvantage the defendant in the majority of cases.

We are left with the task of assessing

$$\text{LR} = \frac{p\left(\bar{X}-\bar{Y} \mid \bar{X},S_x,S_y,C\right)}{p\left(\bar{Y} \mid \bar{X},S_x,S_y\bar{C}\right)} \tag{3.52}$$

Options for the numerator include density functions such as Student's *t*-test and Welch's modification of this.

To overcome the problem of false discrimination because of poor variance estimates, we advocate the use of Welch's modification to the *t*-test. This makes allowance for the variance of the recovered group and the control group to be unequal.

chapter four

Glass found at random and frequency of glass

In Chapters 2 and 3, we looked at the two-stage, as well as the continuous, approach to glass evidence interpretation. The two-stage approach is still the most commonly used and proceeds by testing whether we should discriminate between the samples using various classical significance tests. If, after this stage, we have failed to discriminate between the control and recovered samples, it is necessary that we give a measure of the evidential value of the match. This, we know intuitively, is inversely related to the probability of a chance match given that the suspect is innocent; but as we have seen see in Chapter 3 it is also influenced by other parameters such as the prevalence of glass found at random and the probabilities of transfer and persistence. To estimate these parameters we could, of course, content ourselves with a personal opinion based on experience, but for several reasons it would be preferable to base our assessment on an appropriate body of data. One of these reasons might be the recent ruling in the U.S. suggesting that such expert opinion testimony may not be admissible unless the expert can demonstrate that opinion is indeed scientific and not an art. It is necessary to be precise by what is meant by an "appropriate" body of data. Surely it is a collection that enables us to address the questions that are relevant given the circumstances of the offense and the suspect's alleged involvement.

4.1 Relevant questions

It is much easier to talk in the abstract about "relevant questions" than it is to identify them in a practical situation. The reason for this is simple — it follows from one of the principles of many legal systems that the suspect is under no obligation to offer an explanation for the presence of glass on his clothing. If he *does*, then the questions may become easier to identify. For example, if he says that he had recently been nearby when someone smashed a milk bottle then we could seek a database of milk bottles. This, in itself, is a simplification. The fact that the suspect has offered a possible explanation does not remove his right to the "best" defense. Interpreting this is not trivial, but it might be that the alternative offered should be phrased as, "The glass is from a milk bottle or another source that I cannot identify, unrelated to the crime of which I am innocent." Strictly, in such a case we require a survey of the clothing of persons who have recently broken a milk bottle. Such a specific survey will almost never be available. In general, however, no alternative will have been offered, and we face the task of setting questions which must meet two, possibly conflicting, requirements: they must be relevant, given our limited knowledge of the circumstances, and they must anticipate our ability to answer them. The latter is determined by the availability of data.

4.2 Availability

Creating a good data collection takes time and absorbs precious resources. It has been tempting then, rather than to carry out custom-designed surveys, to use a smaller amount of effort to abstract and organize data which come our way as part of casework. Historically, the most obvious way to do this has been through control data collections though, more recently, valuable data have been collected from clothing encountered in casework.

In the next sections we give a brief historical review of forensic glass data collections. We present the studies done on glass found at random (clothing surveys) and on the distribution of analytical characteristics.

4.3 Glass found at random (clothing surveys)

The presence of glass found at random and the frequency of the analyzed characteristics (generally RI) are two parameters that are related and difficult to separate. They are often both studied in the clothing surveys presented later, but we have chosen to treat them in different sections as we did in Chapter 3. The phrase "clothing survey" will be used to cover any survey which has looked for glass in connection with any part of the "person": so included here are surveys of footwear and hair combings. Two types of populations have been studied: the general population and persons suspected of crime. A summary of the main clothing surveys is given in Table 4.1.

4.3.1 Glass found on the general population

4.3.1.1 Glass recovered on garments

In 1971, Pearson et al.[110] published a survey of glass and paint fragments from suits submitted to a dry cleaning establishment in Reading, England. One hundred sets of jackets and trousers (described as suits) were examined, and debris samples were collected from pockets and pant cuffs. Glass fragments were found in 63 of the suits examined. A total of 551 glass fragments were recovered; 253 fragments, about half of the particles recovered, were

Table 4.1 Summary of Clothing Glass Data Collections

Ref.	Date	Authors	Country	Type of survey	Size
110	1971	Pearson et al.	U.K. (FSS)	Clothing from a dry cleaning establishment	100 suits
111	1977	Davis and DeHaan	U.S.	Men's footwear donated to charity	650 pairs
112	1978	Harrison	U.K. (FSS)	Footwear from casework	99 shoes
105	1985	Harrison et al.	U.K. (FSS)	Clothing from casework	200 cases
107	1987	McQuillan and McCrossan	U.K. (Belfast)	Hair combings from people unconnected with crime	100 people
108	1992	McQuillan and Edgar	U.K. (Belfast)	Clothing from individuals unconnected with crime	432 items of clothing
104	1995	Lambert et al.	U.K. (FSS)	Clothing of people suspected of breaking offenses	589 people
113	1994	Hoefler et al.	Australia	Clothing from individuals unconnected with crime	47 sweatshirts
114	1997	Lau et al.	Canada	Clothing of people unconnected with crime	48 jeans 213 students

found on two suits, with the largest number being 166 fragments on a singlesuit. Results were presented showing the distribution of the glass among the suits and the size of the recovered fragments: about 3% were larger than 1 mm, with the great majority (76%) measuring 0.1 to 0.5 mm. These results are similar to those obtained by McQuillan and Edgar[108] who found that 5% of the fragments recovered in the pockets measured more than 1 mm and 32% more than 0.5 mm (compared to 0 and 6% for particles recovered on the surface).

There was no attempt in Pearson et al.[110] to assess the number of groups of glass present on the suits.

A significant rework of these glass samples, however, was undertaken by Howden.[55] His results showed that "the RI distribution on each suit appeared to be random." This will form an important point in our future discussions. Therefore, it warrants both clarification (the sentence quoted is not quite self-explanatory) and experimental support. Howden remeasured the RIs of available "suits survey" glass fragments (approximately 400 of the original 551 were still available). A group of glass was found to have an RI near 1.5230, which was the same as glass Petri dishes in which the glass had been stored. These fragments were omitted from further work. The remaining glass from each suit was described as random in that it did not form groups of fragments with similar RIs. Howden comments that this is particularly obvious for those suits with large numbers of glass fragments on them. Here, we formalize this as a postulate about glass on clothing of persons unrelated to crime. Specifically, if glass is present it tends to be as small groups, and large groups of glass are rare on such clothing. Both the word large and rare need to be defined; however, this is possible by investigation of the data that has been developed from survey work, and we will make an attempt to do this in this chapter.

With hindsight, a flaw exists in Pearson et al.'s survey. The debris collected was from the pant cuffs, pant pockets, and jacket pockets. Therefore, it appears that debris from the surfaces of the clothing was not collected. This flaw may be of minor significance. Recently, evidence interpretation has emphasized the significance of glass on the surfaces and particularly the surfaces of the upper clothing. Since the suits survey does not examine these, it cannot be used in this way. However, the very significance of the finding of glass on the surfaces is because there is not a lot of glass on the surfaces of clothing. Therefore, the suits survey may have sampled most of the glass that was on the clothing.

In Hoefler et al.'s[113] study on the presence of glass at random on the general population in Australia, only the surface of the garments was examined. The research showed that 41% of the garments had no glass, 53.8% had between one to eight fragments, and surprisingly 5.2% (that is five garments) had more than ten fragments. No attempt was made to assess the number of groups. The majority of the recovered particles were very small, around 0.1 mm. These results corroborate the data provided by Pearson et al.[110] and McQuillan and Edgar.[108]

In 1997 Lau et al.[114] investigated the presence of glass and paint in the general Canadian population. The clothing of 213 students was examined: 1% of upper garments and 3% of lower garments bore fragments on their surface. One fragment was found on five out of six garments, and on the sixth garment two particles were recovered. All fragments were smaller than 1 mm^2, and only one pocket yielded a glass particle. These results differ from the other studies presented previously, as very few fragments were recovered. This may be because of the type of clothing studied (mainly T-shirts), the population, and/or the searching methods.

4.3.1.2 Glass recovered on shoes

Davis and DeHaan[111] examined embedded particulate debris from 650 pairs of men's footwear which had been donated to a charity in Sacramento, CA. Only 20% of the shoes were found to contain colorless glass fragments. The total number of recovered glass fragments was not recorded, but it was noted that of the 261 individual shoes containing glass, 215 contained fewer than four fragments and the remaining 46 contained four or more fragments. Inspection of the histogram for the size distribution of the fragments suggests that a total of about 450 colorless glass fragments were recovered. The Lau et al.[114] survey does not corroborate this study; indeed, on the 164 pairs examined only eight fragments were found on the soles (5% of footwear presented glass). No glass was embedded and no glass was found on the upper part of the shoes. As most particles recovered in Davis and DeHaan[111] were smaller than 0.84 mm and small fragments are difficult to recover without a probe, searching methods may again explain the difference between the two studies. The populations studied were also different.

4.3.1.3 Glass recovered in hair

McQuillan and McCrossan[107] looked for the presence of glass in hair combings (Table 4.2). In samples from 97 friends and relatives of the staff at the Northern Ireland Forensic Science Laboratory, only one glass fragment, smaller than 1 mm, was recovered. No glass was recovered from the hair combings of 20 motor mechanics, and a total of only eight fragments was recovered from a group of six glaziers. No glass was found in the hair combings of two of these six glaziers, one glazier had five fragments coming from four different sources, and the remaining three glaziers had a single fragment (see Table 4.3). These results highlight the evidential value of glass recovered in hair combings.

Table 4.2 Glass Recovered in Hair in Three Different Populations[107]

Type of population	% of persons having glass in their hair
"Normal"	1/97
Mechanics	0/20
Glaziers	4/6

Table 4.3 Glass Recovered in the Hair of Glaziers[107]

Glazier number	Number of fragments	Number of sources
1	0	0
2	0	0
3	5	4
4	1	1
5	1	1
6	1	1

4.3.2 Glass recovered on the suspect population

4.3.2.1 General trends

In 1985 Harrison et al.[105] reported a survey of glass on clothing of persons suspected of involvement in crime. The survey consisted of about 200 cases, and normal U.K. practice was followed with respect to recovery of the glass and selection of fragments for examination. The RI of more than 2000 glass fragments was measured; 40% of these fragments were found to match the respective control samples. Glass was classified as having originated from hair, surface of clothing, pant pockets and cuffs, upper parts of shoes, and embedded in soles. As shown in Table 4.4, the highest proportion of matching glass was found in hair and the lowest proportion was found embedded in shoes.

Table 4.4 Distribution of Recovered Fragments from Harrison et al.[105] and Lambert et al.[104]

Location	Harrison et al.[105] % of total matching	Lambert et al.[104] % of total matching
Hair	69	75
Clothing/surface	54	54
Pockets/cuffs	32	36
Upper parts of shoes	47	43
Embedded in soles	9	N/A

We draw some inferences from this. It is incumbent upon forensic scientists to be fair to the defendant. This might, we feel rightly, be construed to mean that they should not expose him/her to unnecessary risk of random matching. Since most "random" glass is found in the soles of the shoes, this suggests a "top down/outside in" search strategy with the search being terminated when a "large" (yet to be defined) amount of glass has been found. Such a strategy would put hair combings as the first to be searched followed by the surfaces of the upper clothing; next come the surfaces of the pants, shoes, and pockets; and last would be the soles of the shoes. Doing elemental analysis with SEM-EDX, it was found that the majority of the nonmatching fragments large enough to be analyzed did not seem to come from windows.[54,55]

4.3.2.2 *Glass recovered on shoes*

In a previous study, Harrison,[112] following a serious case of criminal damage in Newcastle-upon-Tyne, searched for the presence of glass on 99 shoes (49 pairs and one odd shoe) from ten suspects (therefore, a maximum of ten pairs of shoes can be "guilty" and the remainder should be "innocent"). Debris was collected from the sole, heel, upper parts, inside, and wrappings of each shoe. No glass was contained in 15 of the 99 shoes. In Lambert et al.[104] the number of shoes having no glass was a little lower; on 402 shoes studied, 31% (compared to 15%) had no glass, 12% had matching glass, 23% had matching and nonmatching glass, and 34% had nonmatching glass only. Therefore, it was not unusual to find glass on shoes, and it is worthwhile noting that more than half of the recovered glass fragments did not match the control.

The distribution of group sizes is most like the distribution on glass recovered on pockets or surface: the size of the groups is generally 1 fragment (in more than 80% of the cases).

It is more common to recover glass on a suspect population than on the general population. However, the general findings of both types of surveys corroborate the hypothesis that glass recovered on soles has less evidential value than glass recovered on the surface of clothing.

4.4 Comparison between suspect and general populations: an example

We are now going to concentrate on two surveys. The first survey was carried out on people unconnected with crime in Northern Ireland by McQuillan and Edgar.[108] We will refer to this as ME. The other survey, reported by Lambert et al.,[104] was a large casework study done as a follow-up to that reported by Harrison et al.[105] We will refer to this as LSH. We will explore the extent to which the surveys provide information on the numbers of groups of glass fragments found and the sizes of the groups of fragments.

The ME study was of 432 garments from individuals who had no suspected involvement in crime: members of a youth club, the Ulster Defense Regiment, and the Royal Ulster Constabulary. For each person a pair of garments was examined, i.e., pants with either a jacket or a sweater. They found that 39% of the pant/jacket pairs (including pockets) bore no glass — a proportion which was very close to that reported by Pearson et al.[110] and Hoefler et al.[113] in their surveys.

The LSH survey was set up as follows. During a period of 1984/1985, the six laboratories of the FSS and the Northern Ireland Forensic Science Laboratory collected data on both matching and nonmatching glass that they encountered in searching clothing as part of normal glass casework. Items were collected from 589 individuals, and the RIs of over 4000 fragments were measured. There is a good rationale behind this sort of approach. If, in each of the surveyed cases, we attribute any glass that matches the case's controls as positive evidence for that particular investigation, then any remaining nonmatching glass can credibly be viewed as background material unconnected with any particular incident. Furthermore, without making any contentious assumptions about the guilt or innocence of the casework subjects, it is indisputable that they are people who have come to police notice in connection with the investigation of breaking offenses. Zoro and Fereday[115] have convincingly argued that this is the relevant population to be considered when considering the suspect as an innocent person.

In Figure 4.1 we look at the numbers of groups of glass found per individual and compare the findings of the two surveys. Remember, (1) in the LSH survey we consider only the nonmatching glass, and (2) we are considering surface fragments found per individual, i.e., found on an upper/lower pair of garments. The proportion of garment pairs with no glass on their surface was higher in the ME survey (64% compared with 42%). Otherwise, the two distributions have similar shapes. In the LSH survey the groups are smaller than in the ME survey (see Figure 4.2).

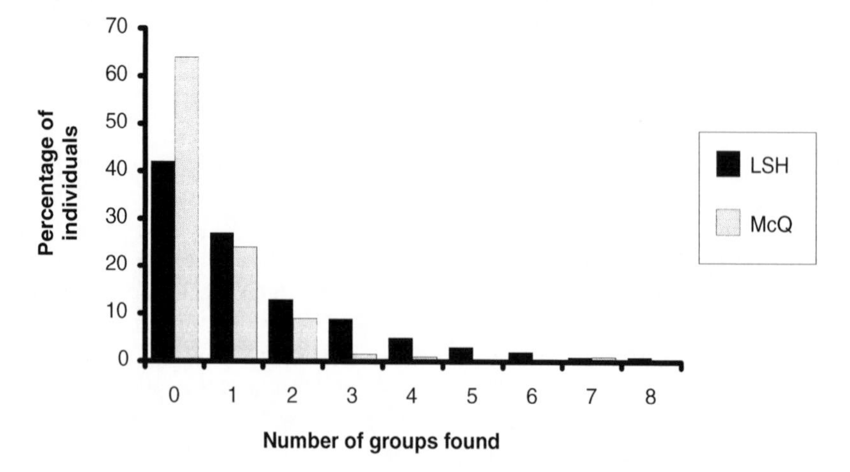

Figure 4.1 Number of groups of glass found per individual (surface only) in the LSH and McQ surveys.

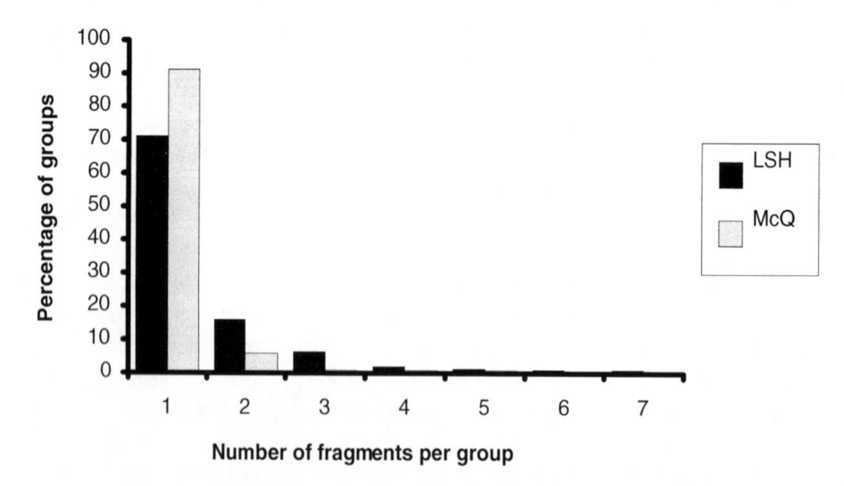

Figure 4.2 Distribution of the size of groups of fragments in the LSH and McQ surveys.

4.5 Estimation of the probability of finding at random *i* groups of *j* fragments

Tables 4.5 and 4.6 give a summary of the data from the ME survey in a form that can be used in the type of calculations we are advocating — that we intend to act as a resource in the practical application of these statistical methods. It would be preferable if each country could perform its own survey; however, we realize the difficulties and costs in this. The ME data has been summarized for size of fragments into small groups (one to two fragments) and large groups (three or more fragments), as this was the thinking at the time. The choice of the definition of small and large is arbitrary and subjective. We hope that a study of the survey data by other investigators might lead to similar conclusions. At that time our standard practice was to ask people to subjectively estimate the probability of transfer of a large or small group of glass. These coarse divisions were about all that could be handled for transfer estimation at that time, and it was necessary to align the S terms in a similar way. It was not until the development of the graphical models (vis-à-vis TFER, a program to assemble estimate transfer probabilities from data, discussed in Chapter 5) that it was realistic to try to model the transfer of 0, 1, 2,..., etc. glass fragments.

Table 4.5 The Number of Groups of Glass Found for Different Search Strategies from the ME Survey

Number of groups of glass	Upper garments surface only	Upper and lower garments surface only	Upper garments surface and pockets	Upper and lower garments surface and pockets
P0	0.811	0.636	0.641	0.403
P1	0.146	0.238	0.180	0.272
P2	0.029	0.087	0.053	0.087
P3	0.000	0.010	0.063	0.053
P4	0.010	0.010	0.024	0.092
P5	0.005	0.005	0.015	0.015
P6	0.000	0.000	0.000	0.019
P7	0.000	0.005	0.000	0.005
P8	0.000	0.000	0.015	0.019
P9	0.000	0.000	0.005	0.015

Note: Data have been grouped using the Evett and Lambert grouping algorithm. Data were reworked by Buckleton and Pinchin from the raw data and differ slightly from the published set.

Table 4.6 The Size of Groups of Glass Found for Different Search Strategies from the ME Survey

Size of groups of glass	Upper garments surface only	Upper and lower garments surface only	Upper garments surface and pockets	Upper and lower garments surface and pockets
1 or 2 fragments	0.980	0.971	0.958	0.965
3 or more	0.020	0.029	0.042	0.035

Note: Data have been grouped using the Evett and Lambert grouping algorithm. Data were reworked by Buckleton and Pinchin from the raw data and differ slightly from the published set.

The data in the large and small format are not as directly useful when assessing, say, S_6 as they could be, but they are useful adjuncts to subjective judgment.

4.6 Frequency of the analyzed characteristics

At the beginning of this chapter we mentioned that most studies on glass found at random have also considered the frequency of physical and/or chemical characteristics of the recovered glass. This is very important because the frequency of the analyzed characteristics is of interest only if the glass is present by chance and, therefore, does not come from the control. This conclusion follows logically from the Bayesian approach, but only more recently has been accepted. The first databases used to estimate the frequency of the analytical characteristics included mainly glass coming from controls. Tables 4.7 and 4.8 show a summary of the studies that will be presented.

Table 4.7 Summary of Control Glass Data Collections

Ref.	Date	Authors	Country	Type of survey	Size
	1968	Cobb	U.K. (FSS)	Control glass from breaking cases in the West Midlands	175 controls
25	1972	Dabbs and Pearson	U.K. (FSS)	Fire survey data	939 samples
116	1977 1984	Lambert and Evett	U.K. (FSS)	Control glass from casework	9000+ controls
16	1982	Miller	U.S. (FBI)	Flat glass from casework	1200 samples
19	1986	Buckleton et al.	New Zealand	Glass samples from 388 vehicles	513 samples
117	1989	Faulkner and Underhill	U.K. (MPFSL)	Samples of nonwindow glass	300 samples

Table 4.8 Summary of Clothing Glass Data Collections Where Analytical Data Were Collected

Ref.	Date	Authors	Country	Type of survey	Size
110	1971	Pearson et al.	U.K. (FSS)	Clothing from a dry cleaning establishment	100 suits
111	1977	Davis and DeHaan	U.S.	Men's footwear donated to charity	650 pairs
112	1978	Harrison	U.K. (FSS)	Footwear from casework	99 shoes
105	1985	Harrison et al.	U.K. (FSS)	Clothing from casework	200 cases
108	1992	McQuillan and Edgar	U.K. (Belfast)	Clothing from individuals unconnected with crime	432 items of clothing
104	1995	Lambert et al.	U.K. (FSS)	Clothing of people suspected of breaking offenses	589 people

4.7 Control glass data collections

Initiatives to collect control glass examined in the course of routine casework started in the 1960s and are now widespread. The information on the RI distribution of control glass (information on elemental analysis is scarce) shows that the frequency of the RI depends on the type of glass, its age, and its geographical origin.[19,86,91]

Imagine that the examiner finds a group of fragments on the suspect's clothing that "match" the control from the scene. Let us assume that the only collection available to us is a control data collection. In that case we are driven to ask the following kinds of questions.

What is the probability that a group of fragments from some other glass previously submitted would have matched the recovered fragments?

This is the "coincidence probability" approach that we talked about in Chapter 2. But, of course, we have problems straight away: what do we mean by "some other glass source"? If we can break our data collection into classifications such as "window," "container," etc., then which of these should we use? As we have seen in Chapter 3 the frequency is only of interest if the suspect did not break the window (or any other control object), and in that case there is no justification for assuming that the recovered glass must have come from another window. Resolution of this apparent conundrum was never achieved by any kind of theoretical consideration. Instead, it became one of the main drivers of the move in the late 1970s/early 1980s to routine use of elemental analysis of glass, largely because of its potential for classifying glass fragments, and, similarly, it was one of the motivations for the introduction of the interferometer.[32,33]

Let us assume that elemental analysis shows that the recovered fragments are, indeed, window glass. Then we can consult our control window glass data and make the following statement.

There is a certain probability that the recovered fragments would match a different window control.

If we assume that the control glass data collection is representative of the distribution among broken windows in general, then this statement undoubtedly has some value. The smaller the probability of such a coincidental match is, the stronger the evidence.

If there is no data suggesting that the recovered glass may have come from a window pane, it would be erroneous to use control databases. Indeed, clothing surveys, as well as miscellaneous studies, have shown that glass broken at random originates mainly from sources other that windows. For example, Zoro and Fereday[115] reported the analysis of responses from a random sample of 5000 members of the general public on their recent contact with breaking glass. Some general conclusions are that about a third of respondents claimed to have broken or been close to breaking glass within the previous seven days. Of these, most had only been in contact with one source of broken glass, and most of the broken glass objects were not windows.

The same year Walsh and Buckleton[118] reported a survey of glass fragments found on footpaths in different areas of Auckland, New Zealand. The main aim of this survey was to estimate the proportions of the different types of glass encountered. From 52 km of footpath, 1068 pieces of colorless glass were collected. Footpaths were classified as being in central city areas, residential areas, light industrial areas, and areas where buildings were absent. Container glass dominated in all areas, totaling about 70% of the glass collected. Building windows represented about 12%, and most of the remaining glass was from vehicles. Even in the city areas only about 19% of the glass was from buildings.

Control data collections only allow part of the job to be done. It constrains us to addressing a question that may not be the most relevant one as far as the deliberations of a court are concerned. This has been recognized for as long as data collections have been in existence. It seems reasonable to take this body of data as a suggestion that control glass surveys are not correctly balanced with regard to the type of glass, and that clothing surveys should be used whenever possible.

4.8 Clothing surveys

In most clothing surveys presented earlier the physical characteristics of the recovered fragments have been analyzed in order to see if the RI distributions of glass recovered at random and control glass were different. Only occasionally has elemental analysis been used in order to clarify any putative classification of the recovered particles.

4.9 Characteristics of glass found on the general population

4.9.1 Glass recovered on garments

Pearson et al.[110] measured the RI of 551 fragments of glass recovered on suits (see Table 4.8). The RI distribution was clearly different from that of the Cobb data (see Table 4.7). Thus, the tentative conclusion expressed by the authors was that about 30% of the glass was probably window glass. Howden[55] studied this hypothesis doing elemental analysis on the fragments recovered in Pearson et al.[110] It was shown that most glass on clothing is not window glass, but is composed significantly of tableware. There are important consequences if this hypothesis is true, and, therefore, it is worthwhile offering some experimental support for it. Howden reexamined suitable and overlapping subsets of the fragments using various elemental techniques. He found that

> Consideration of the refractive indices, atomic absorption, X-ray fluorescence and microprobe analysis results all lead to a similar overall conclusion. Namely that much of the glass on clothing originates from tableware or colored glasses and that not much comes from windows or colorless bottles.

If this latter hypothesis is true, then the presence of matching window glass in the population studied, that is glass recovered on clothing that can be shown to be window glass that matches a window control, may be very evidential because window glass on clothing is rare. Such an interpretation would be valuable and is worthwhile investigating and, if possible, quantifying. McQuillan and Edgar's study shows comparable results.

4.9.2 Glass recovered on shoes

Davis and DeHaan[111] measured the RI of 366 of the larger glass fragments recovered in their survey. The RI distribution showed that a considerable number of fragments had an RI greater than 1.533. The authors concluded that most of the glass came either from containers or window panes. In the study mentioned previously, Harrison[112] analyzed the RI of 363 glass fragments recovered from 62 of the 77 shoes. The RI distribution was compared with and found to be markedly different from the available control glass data. Comparisons between the RI distribution of the glass recovered on the shoes and the subsets of glass in the Home Office Central Research Establishment control data files showed that the best fit is to "container" glass. This comparison also suggests that the shoes had more glass with low RI (less than 1.516). This deficiency could be made up by increasing the contribution of "Container glass [sic] which is not bottle." It seems plausible to conclude from this comparison that the glass on the shoes was largely not of window origin, but was most likely composed of container glass with a significant contribution of tableware. This hypothesis is partly supported by Walsh and Buckleton's[118] results.

Comparisons between Davis and DeHaan's[111] study in the U.S. and Harrison et al.'s study[105,112] in the U.K. show the same general tendencies. The difference in the RI distribution, that has also been shown for control glass, is explainable by the fact that the glasses found in the two countries are likely to have come from different manufacturers and plants. The RI distribution also depends on the date of the studies and is the plausible explanation for the differences between Pearson et al.[110] and Harrison et al.[105]

If we accept Harrison's point that "clothing" surveys should be used and we observe that control collections are dominated by window glass and further we accept that glass on clothing is largely container glass dominated by tableware, then again we see that control glass collections are of limited use, except if a method has been used to classify glass. Even then, they will answer only part of the question.

4.10 Characteristics of glass found on the suspect population

In their study on glass found at random on persons suspected of crime, Harrison et al.[105] produced RI frequency distributions for the raw and grouped nonmatching glass data. These were compared with the published control glass data and shown to be generally different. It appeared that a considerable proportion of the nonmatching glass originated from sources other than windows. For example, the RI distribution of glass embedded in the soles of footwear agreed with the RI distribution of control container glass. A very limited amount of elemental analysis on the nonmatching recovered glass suggested that about a third of the glass in the RI range typical of windows had originated from a nonwindow source. The RI distribution of the matching glass in the survey was, not surprisingly, similar to that of the control glass data.

It is tempting to argue that even some of the "nonmatching glass" is actually matching and, therefore, of largely window origin. This allegation could be made because of the propensity of grouping criteria to "drop-off" outliers. If this were true then it would suggest that the estimate of Harrison et al. of the fraction of window glass in their nonmatching survey is an overestimate. However, J. Lambert (personal communication, 1995) has convincingly replied to this argument by showing that the distribution of nonmatching glass on people with only nonmatching glass is not different from the distribution of nonmatching glass on people who also have matching glass. In view of this, it seems reasonable to accept Harrison et al.'s conclusion.

4.11 Comparison between suspect and general populations: an example

Figure 4.3 shows the distribution of RI measurements for the groups of nonmatching fragments in the LSH survey. The ME distribution is not shown in detail, but in Figure 4.4 we compare the general shape of the two distributions. There is reasonable agreement between the two. The higher proportion of LSH data in the range of 1.5150 to 1.5180 suggests a greater proportion of window glass than in the ME survey. Since window glass is the most common type to be broken during criminal activity, and since police tend to arrest suspects who they believe engage in regular criminal activity, it is to be expected that clothing obtained from such suspects would contain more window glass than the clothing of members of the public who have no involvement in crime.

However, the imagined need to discriminate between different types of glass is not so immediate when we have a clothing survey to refer to.

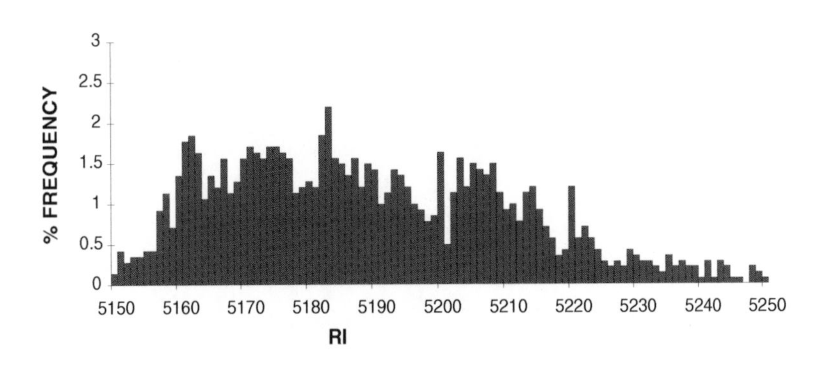

Figure 4.3 RI distribution for groups of nonmatching glass in the LSH survey. (The RIs have been transformed by subtracting one and multiplying by 10,000.)

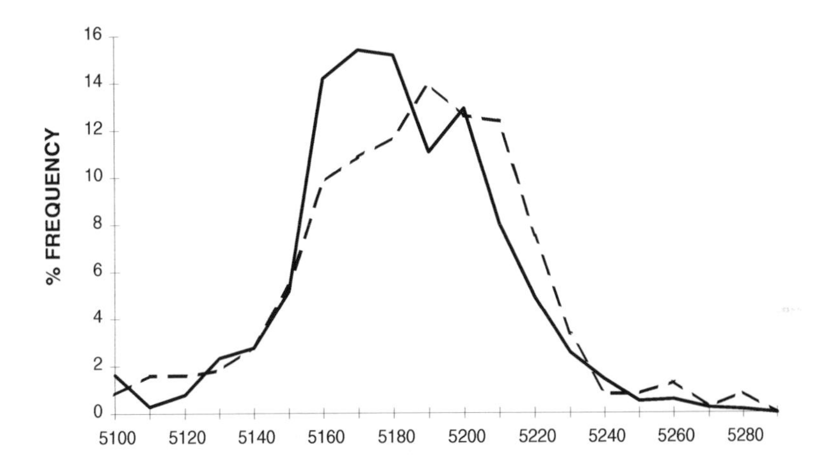

Figure 4.4 Comparison of the RI distribution for nonmatching glass in the LSH (———) and ME (— — —) surveys.

4.12 Summary

In conclusion, we would like to make a "philosophical" point. When asked to fit a simple geometric shape to the "glass recovered from shoes" distribution or many other "glass from clothing" distributions, most glass analysts choose a rectangle. The logical consequence of this "thought experiment" is that most glass analysts feel that a reasonable approximation is to treat frequency as a constant. This is of interest because so much effort has been put into assessing frequency, and it suggests that we could usefully turn our attention elsewhere. Where? We would suggest that assessing the amount of glass on persons and the probabilities of transfer are useful places for this investment.

chapter five

Transfer and persistence studies

As we have seen in Chapter 3, background information such as transfer and persistence is essential to evaluate glass evidence, and its importance cannot be overstated. It is very important that the glass examiner be aware of the data that have been published on the subject. If a search is performed, this data is necessary to form an opinion on the value of evidence. Transfer and persistence probabilities are very difficult for the expert to assess as they may depend on various factors. The amount of glass expected to be recovered is reported to be influenced by the type and thickness of the glass, the distance at which the window has been broken, the garments of the suspect, the time elapsed between the incident and the arrest, and the activities of the suspect between the offense and the arrest. It may also be affected by other variables such as the size of the window, whether or not entry was gained to the premises, or the weather at the time of the incident.

Such information can assist in decision making. For example, given the information supplied on the case submission form, would I expect to find glass on the clothing of this suspect, and if so, where and how much? The answers to these questions should guide the forensic scientist as to which item should be examined first and which items should not be examined at all. A survey of data from submission forms actually suggests that the data in them may be of little use, which is a disappointing result for those advocating a rational approach to glass casework.[119]

The data on transfer is extensive. However, it is quite difficult to organize due to a very large number of variables. This also makes it a difficult task for the examiner to estimate transfer probabilities in any specific case. In order to facilitate this, we introduce TFER, a program to assemble an estimate from the data.

5.1 Transfer of glass

We begin by observing that most glass cases involve flexible glass (relatively thin panes) that has been broken in flexion. It has been previously noted that the nature of the impacting object is not crucial to breakage under flexion. It must have sufficient kinetic energy, but beyond that the exact nature of the impacting object is thought to be, at least, secondary. Breakage is thought to progress by the extension of radial cracks and by the propagation of concentric cracks. Later in this chapter we will discuss the preponderance of surface fragments in glass backscattered from breakage. Plausible hypotheses are available for this. An examination of glass pieces after breakage will reveal many fragments missing from the edges of the cracks. It is reasonable to believe that these fragments have been ejected during propagation of the cracks by tension or compression forces.

Kirk[120] first described the transfer of fragments onto the garments of a person breaking a window, but it was not until 1967 that Nelson and Revell[2] demonstrated the phenomenon scientifically by publishing photographs of glass particles projected backward. Since then the mechanisms of transfer and persistence have received some attention. The studies described in the following sections provide, in general terms, three types of information to assist in assessing the effects of the previous factors.

1. Glass smashing experiments, which examine the number and distribution of glass fragments transferred to the ground, involve smashing different types of glass with different methods of breakage.
2. Glass smashing experiments that look at quantity of glass fragments transferred to individuals who are nearby. Work on the influence of parameters such as the weather or the breaking device, as well as research on secondary transfer, is included in this group.
3. Studies on the retention of glass fragments on clothing with respect to time and subsequent activity.

5.1.1 *Transfer of glass to the ground*

5.1.1.1 *Number, size, and distribution of the fragments*

Because of the complexity of the transfer process, it is necessary to control as many parameters as possible. Nelson and Revell,[2] Locke and Unikowski,[121,122] and Locke and Scranage[30] proposed standardizing the breakage of windows with a pendulum.

Nelson and Revell's study[2] was performed on 19 panes of 0.6 × 0.9 m glass. Two thicknesses (3 and 6 mm) were used, and the panes were broken either with a pendulum or a hammer. The results of their research are as follows.

1. Fragments were projected as far as 3.3 m.
2. A stroke in the middle of a pane produced more fragments than a stroke in a corner.
3. The faster the pendulum, the smaller the hole and the fewer fragments were produced.

4. If the speed of the pendulum was high, then the dispersion (distance, angle) of the fragments was reduced.
5. Although most of the backscattered fragments from 3 mm window glass were flakes and chips, the fragmentation from 6 mm glass also included an appreciable number of needle-like slivers.
6. Transfer is a variable phenomenon, especially when the windows are broken with a hammer.

Their final conclusion was

> In any investigation where window glass has been broken it is well worthwhile — in fact almost a matter of duty — for the investigator to secure samples of the broken glass and the clothing of the suspect and submit them for laboratory examination.

This research revealed a number of patterns. However, Nelson and Revell performed only a small number of experiments and did not give quantitative data regarding the number and the size of the fragments transferred. Locke and Unikowski[121,122] and Locke and Scranage[30] have carried out quantitative studies on transfer. The aim of the first study was to establish the reproducibility of transfer and give an idea as to how many fragments are transferred depending on the distance and the position of the person standing nearby. The apparatus used to break windows was standardized, and the debris was collected in trays according to distance and direction. The eight experiments demonstrated that transfer is a variable phenomenon even when the conditions are standardized to a maximum. Nevertheless, it was possible to show the following.

1. The number of fragments varied by a factor of 4.
2. This number depends on the distance and declined very rapidly.
3. Fragments smaller than 1 mm were homogeneously distributed in the different sectors, but bigger fragments were mostly found in the inner sectors.

Pounds and Smalldon,[123] who employed less standard breaking procedure (the windows were broken with a hammer or a brick), obtained similar results to Nelson and Revell (point 1) and Locke and Unikowski (points 2 and 3). Luce et al.[124] also confirmed point 2 (Locke and Unikowski): the number of fragments observed in their research declined by a factor of 4 to 5 for every 45 cm.

5.1.1.2 Influence of the window type and size

Having demonstrated that patterns could be revealed even in the face of varying experimental conditions, Locke and Unikowski[122] studied the effect of size and thickness of windows, as well as the influence of the type of glass, on the number of fragments transferred. The characteristics of the 24 windowpanes are shown in Table 5.1.

Table 5.1 Types of Window Tested in Locke and Unikowski[122]

Size	Thickness	Wired	Patterned
1 m × 1 m × 4 mm	1 m × 1 m × 4 mm	1 m × 1 m × 6 mm	1 m × 1 m × 4 mm
0.5 m × 0.5 m × 4 mm	1 m × 1 m × 6 mm	0.5 m × 0.5 m × 6 mm	0.5 m × 0.5 m × 4 mm
0.25 m × 0.25 m × 4 mm	1 m × 1 m × 10 mm	0.25 m × 0.25 m × 6 mm	0.25 m × 0.25 m × 4 mm

The experiments, duplicated for each type of glass, showed that the number of fragments does not depend on the size of the windowpane, but rather on the total amount of cracking and the degree of disintegration. Whether the window was patterned or not did not influence the number of projected particles, but if the windowpane was wired or very thick, more fragments were produced. This phenomenon could be explained by the fact that more energy was needed to break wired or very thick glass, and, therefore, more energy was dissipated. This hypothesis was corroborated by the observations of Luce et al.[124] who suggested that force and number of particles are related. However, Nelson and Revell's[2] conclusions on the correlation between the number of fragments and the speed of the pendulum do not necessarily support this hypothesis. In the case of wired glass, another explanation would be that the particles are not only induced by radial and concentric fractures, but also by fractures between the wire and the glass.

5.1.1.3 Presence of an original surface

As the presence of an original surface on a recovered glass fragment can establish its origin (container or flat glass) and also influences its RI, it is helpful to determine if fragments projected backward originate from the interior, the exterior, or the bulk of the window. Following these observations, Locke and Scranage[30] painted windows with black and red ink and demonstrated that about 50% of the fragments measuring 1 to 0.25 mm came from the surfaces. The exterior to interior surface ratio can vary from 3:1 to 30:1 depending on the experiment. The ratio also depended on distance: the farther the person was from the window, the greater the fraction of fragments that were projected from the rear. This could have been due to the fact that secondary breaking is the most likely origin of fragments projected at a distance. Since these fragments have the same chance of coming from the interior or the exterior, the ratio is reduced (it tends toward 1:1). For particles projected far from the window, a small number of particles is sufficient to change the ratio. Locke and Scranage,[30] Zoro,[125] and Luce et al.[124] obtain comparable results. In experiments on glass transferred onto garments, this tendency was also confirmed: indeed, Allen and others[126-129] have shown that 60% of the fragments had an original surface, most of them coming from the front of the pane.

5.1.2 Transfer of glass broken with a firearm

Francis[130] studied the backward fragmentation of glass as a result of an impact from a firearm projectile. Three types of glass (ordinary clear house glass, 3 mm thick; frosted wired glass, 5 mm thick; and clear laminated glass, 6.4 mm thick) and eight firearms and projectiles were tested. The panes were broken at a distance of 4.0 m, with the shooter being directly in front of the window. The fragments present on a grid measuring 4 × 4 m were then counted in a manner similar to that employed by Locke and Unikowski[121,122] or Luce et al.[124] A large number of fragments were recovered; therefore, the author recommended that "should a shooting occur through a window, the suspect's clothing, shoes and hair should be examined for the presence of glass window." No result has been given for the number of fragments recovered on the shooter. However, current research (T. Hicks, personal communication, 1999) is being performed in this area.

5.1.3 Transfer of vehicle glass

Locke et al.[131] and Allen et al.[132] have studied the transfer of vehicle glass. In the first study, the authors investigated the nature of the fragmentation process and the total number of fragments generated (tens of thousands of casework size particles). In the second study, the authors examined the distribution and number of glass particles transferred when breaking toughened and laminated screens. When the window was crazed, or had fracture lines without significant breakage, over 1400 fragments in the range of 0.25 to 1.0 mm were found in the car. Although there was considerable variation, it was shown that more particles were produced by shattering than by crazing. By examining the spatial distribution, it was found that very few fragments were present at distances greater than 1.5 m and at the sides of the car.

5.1.4 Transfer of glass to individuals standing nearby

The work presented previously takes into consideration the number of fragments when distance, position, and glass type are varied. However, it does not give an estimate of the number of fragments that can be projected onto a person. In addition, it is uncertain whether the observations made on the particles transferred to the ground are also applicable to fragments transferred to garments or hair.

The transfer of several types of glass (window, windscreens, and broken glass) onto several receptors (garments, hair, and shoes) is presented next.

5.1.5 Transfer of window glass to garments

In order to standardize as many parameters as possible, Allen and Scranage (5 experiments),[126] Pounds and Smalldon (7 experiments),[123] and Luce et al. (8 experiments)[124] have studied the transfer of glass when breaking a window with a pendulum. In order to estimate the number of fragments that may be transferred in actual breaking, experiments have also been carried out on glass broken with a tool or a stone.

5.1.6 Transfer of glass with a pendulum

Allen and Scranage[126] performed five experiments breaking a 1 m × 1 m × 4 mm window with a pendulum. They found that the number of fragments transferred to garments* was about 28 fragments at a distance of 0.5 m, 5 fragments at 1 m, 1 fragment at 1.5 m, and 2 fragments at 3 m. Most of the fragments were found on the sweater, but substantial numbers were also found on the socks. However, with greater distance, more fragments were recovered on socks than on the sweater. Very few particles were found on the pants. One windowpane was broken by a pendulum with a glove attached in order to see how many fragments may be transferred to the glove: 20 glass fragments were found embedded. Of these fragments, 18 were larger than 0.5 mm and, perhaps contrary to expectation, came predominantly from the rear surface.

The results of Luce et al.[124] (8 experiments) show that the number of transferred fragments varied between 19 and 14 recovered from the pants, shoes, sweatshirt, and hair of a person standing at 0.45 m and 9 and 3 for a person standing at a distance of 0.90 m. The variability of the results is not surprising, given the variability in the number of fragments transferred to the ground. But even if the variability is high, this research shows that fragments are transferred to persons standing nearby and allows a comparison of results with the studies of glass transferred to the ground.

5.1.7 Glass broken under conditions similar to casework

In casework modus operandi is often unknown. Subsequently, many practitioners think that if it is unknown, then it is not possible to estimate the number of fragments likely to have been transferred. We have found that when subjects are asked to stand in a natural position and strike a window, the positions taken are remarkably similar. Therefore, we believe that when breaking a window the breaker can stand in three positions:

- He/she can charge the windowpane with the shoulder or the entire body; the distance would then be 0.0 m.
- The breaker can strike the window pane with a tool or the foot; the distance would then be approximately 0.6 to 0.9 m.
- He/she may throw an object; the distance would then exceed 1.5 m.

Therefore, we argue that even if the exact distance between the offender and the window is unknown, it is possible to make some reasonable estimate. The experiments presented next simulate the last two breaking procedures.

* Acrylic hat, polyester and wool socks, woolen sweater, synthetic pants, and leather gloves.

Luce et al. (4 experiments),[124] Pounds and Smalldon (4 experiments),[123] Hoefler et al. (20 experiments),[113] Hicks et al. (22 experiments, 15 breakings with a hammer and 7 with a stone),[133] Allen et al. (15 experiments),[132] and Cox et al. (15 experiments; personal communication, 1996) have studied the transfer of glass resulting from the breakage of a window with a tool or a brick. The aim of these studies was to determine if the number of fragments transferred would differ from those obtained in the experiments with a pendulum. Their results are summarized in Table 5.2.

Because of the different experimental conditions, it is difficult to compare these results. First, different types of clothing have been used. For example, Hoefler et al. have used 20 different jeans and sweatshirts, and Hicks et al. have used a cotton tracksuit (person standing at 0.6 m) and one coarse pullover and a pair of jeans (person standing at 0.8 m) throughout all experiments. Allen et al. and Cox et al. have used three types of clothing (shellsuits, tracksuits, and a sweater and jeans). The size of the panes was also varied. For example, the window panes used were 60 cm × 60 cm × 3 mm in Hicks et al., a set of four panes having the dimensions 25 cm × 36 cm × 2.2 mm in Hoefler et al, panes of different dimensions in Allen et al., and panes measuring 48 cm × 59 cm × 4 mm in Cox et al. Nevertheless, a study of Table 5.2 shows that more fragments are transferred with the hammer than with the pendulum; the number of particles decreases with distance; and, according to Cox et al., the number of fragments initially transferred does not depend (for the clothing used) on the type of garment. One can also observe, with the most recent studies which replicate the experiments 15 or 20 times, the high variability in the number of fragments transferred. As shown in Table 5.2 this number was on average higher in Hicks et al.[133] This difference may be due to the breaking procedures, the force, the size of the panes, or/and the natural variation. According to Hicks et al.[133] and Luce et al.,[124] on average more fragments would be produced when breaking the windowpane with multiple blows. The recovery procedure is also important and may explain some differences. In all studies the fragments were recovered by shaking, as recommended by Pounds,[134] then separated into size categories and counted.

The way the subjects disrobed may have influenced the results. Hoefler et al.[113] and Allen et al.[126-129] have shown that when disrobing, some glass was lost (the range of fragments lost during disrobing was one to eight particles in Allen et al.[126-129]). However, if we include the results of Cox et al. on transfer of glass on upper and lower garments and compare all the studies done on transfer, which of necessity are performed under disparate conditions, the most likely explanation for the variance is the inherent variability of the transfer process in Figure 5.1.

Hicks et al. showed that 3 to 10% of the particles transferred were larger than 0.5 mm. Luce et al.[124] confirmed this result.

As a practical consequence of the studies, it is important, in casework, to consider not only submitting clothing for examination, but also the sheets on which the suspect has disrobed.

Table 5.2 Summary of Transfer Experiments

Distance	Repeats	Research	Breaking device	Top			Pants		
				Ret[na]	Mean	Range	Ret[n]	Mean	Range
0.45 m	1	Pounds and Smalldon	Hammer	—	27	—	—	—	—
	2	Luce et al.	Hammer (1 strike)	M	12	7–13	M	2	1–3
	3	Luce et al.	Hammer (>1 strike)	M	50	12–72	M	3	1–3
0.50–0.80 m	20	Hoefler et al.	Jemmy bar	M	12	3–24	M	9	3–22
0.50 m	5	Cox et al.	Crowbar (1 strike)	L	39	22–80	L	27	13–38
	5			M	27	14–37	M	27	16–41
	5			M	26	12–46	M	25	14–36
		Allen et al.	Crowbar (1 strike)	M	12	6–28	M	9	5–15
				L	16	5–30	L	16	6–29
				M	15	10–23	M	13	8–29
0.60 m	15	Hicks et al.	Hammer (>1 strike)	M	127	44–241	M	31	5–81
0.75 m	1	Pounds and Smalldon	Hammer	—	11				—
0.80 m	15	Hicks et al.	Hammer (>1 strike)	H	40	15–72	M	24	4–85
1.35 m	1	Pounds and Smalldon	Brick	—	4				—
1.50 m	7	Hicks et al.	Stone	M	4	0–16	M	4	0–10
	7			H	5	1–14	M	6	0–16
1.95 m	1	Pounds and Smalldon	Brick	—	3				—

[a] Retention.

Note: The retention characteristics are described as low (L), medium (M), and high (H).

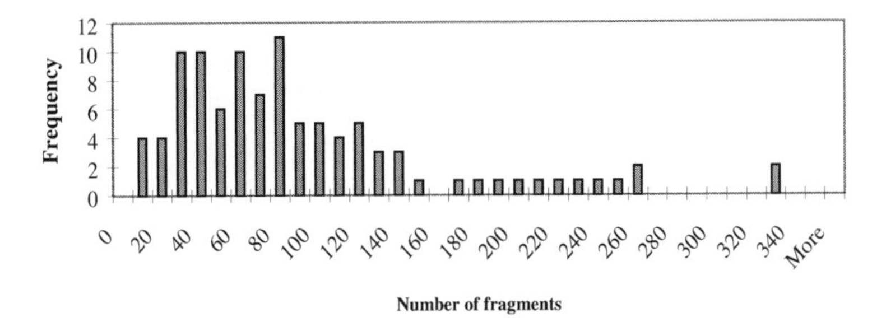

Figure 5.1 Number of fragments transferred on lower and upper garments at T = 0 when breaking a windowpane with a tool. (Results of studies by Hoefler et al., Hicks et al., and Cox et al.)

5.1.8 Transfer of vehicle glass and absence of glass

Rankin et al. (personal communication, 1989) have studied the transfer of vehicle glass onto garments (the authors were investigating whether the absence of glass could be an indication that a person had not smashed a window). As a relatively small number of glass fragments were found on the leather jacket used when reconstructing the case, it was concluded that it was possible that no glass would be found given the case circumstances. This is confirmed by the research done on transfer of windowpane glass, where it has been shown that sometimes very little glass is transferred. Therefore, the absence of glass is not an absolute indication that the suspect has not committed the crime.* However, even this negative evidence warrants careful interpretation.

5.1.9 Transfer of glass when a person enters through a window

Allen et al.[127] have studied the transfer of glass when breaking and stepping through a windowpane. Three types of garments (shellsuit, tracksuit, and a sweater and jeans) were used. It was found that rubbing the subject's sleeve against the broken edge of the pane had little effect on the number of fragments transferred.

5.1.10 Influence of the weather on transfer

Allen et al.[127] also investigated the influence of wetness of the clothing on transfer using the following procedure: "The window was broken and the dry clothes taken immediately. Fresh clothes were wetted with a hand

* This is a very general hypothesis. However, as seen in Chapter 3, which is devoted to the Bayesian approach, there may be other hypotheses.

sprayer, until the clothes were damp to touch. Another window was broken (with a crowbar at a distance of 0.5 m) and clothing removed." The conclusion of their study was that "Wet clothing (in particular shellsuits) may retain larger numbers of fragments and a higher proportion of fragments greater than 0.5 mm as compared to dry clothing." However, as the number of fragments is highly variable, this result must be taken with caution and further research is necessary to confirm this result.

5.1.11 Transfer of broken glass

As mentioned earlier, Underhill[135] compared glass transferred to clothing during the breaking of a window with that transferred during clearing up previously broken glass. More fragments (typically more than 100) were transferred during clearing up, but a smaller proportion of these were found to have original surfaces. Therefore, a high number of outer surface fragments would suggest backscatter. The relationship between fragment shape (needle, chunk, and flake) and the mode of acquisition was investigated; however, no significant correlation was found.

5.1.12 Transfer of window glass to hair

Several authors have studied the transfer of glass to hair or headwear. Glass was either transferred by breaking a window with a pendulum, a tool, or a brick or when clearing up the glass. As can be seen in Table 5.3, glass was not always transferred, and, when there was transfer, the number was variable. However, again, these findings give an order of magnitude to the number of glass fragments that one would expect to recover from hair a few minutes after the breaking.

Table 5.3 Number of Fragments Recovered in Hair or Headwear

Approximate distance	Research	Breaking device	Number of experiments	Headwear or hair	
				Mean #	Range
50 cm	Pounds and Smalldon	Hammer	1	3	—
	Luce et al.	Hammer (1 strike)	4	2	0–6
	Luce et al.	Pendulum	2	1	1
	Luce et al.	Hammer (>1 strike)	4	7	1–16
80 cm	Pounds and Smalldon	Hammer	1	3	—
100 cm	Luce et al.	Pendulum	2	1	1
	Allen et al.	Crowbar	10	2	0–7
140 cm	Pounds and Smalldon	Brick	1	0	—
200 cm	Pounds and Smalldon	Brick	1	0	—

Underhill[135] also reported the average number of fragments that were recovered in hair after 10 minutes. Several experiments performed were described as training exercises, backward fragmentation, and clearing-up experiments. The windowpanes were broken at a distance of 0.60 m with a crowbar. On average, the number of glass particles recovered was two in clearing-up experiments and one when breaking the window. At most, 20 fragments were recovered in the hair (training exercises).

5.1.13 Transfer of window glass to footwear

No study has been published on glass transfer by kicking of the pane, which is an obvious omission given the frequency with which items are submitted from this modus operandi. Luce et al.[124] have reported the number of glass fragments that were transferred to footwear when breaking a windowpane. On running shoes, 8, 0, 8, and 13 fragments were recovered when breaking a windowpane with a single strike at a distance of about 0.50 m. In the four experiments involving multiple hammer strikes, 14, 10, 22, and 47 particles were recovered. An interesting finding was that more glass was found on footwear than on pants or upper clothing when the person was standing 1.0 m from the breaking window. This result has been confirmed by Allen and Scranage.[126]

5.1.14 Secondary and tertiary transfer

Bone[136] describes secondary transfer experiments between two individuals wearing woolen sweaters with very good retentive properties. Up to 500 glass fragments were placed on the front of the donor, who then hugged the recipient for 5 to 60 seconds. Donor and recipient garments were then removed and searched. In 12 experiments, the number of glass fragments remaining on the recipient sweater after contact ranged from 0 to 9. A high, though variable, proportion of glass was lost from the donor garment when it was removed. In general, more fragments dropped off the recipient sweater when it was removed than were retained on it. The contact time did not have much effect on the percentage of glass remaining on the donor sweater, but it did affect the number found on the recipient.

Bone's research did not actually transfer glass by breaking a window. Holcroft and Shearer[137] conducted experiments where glass was transferred to individuals by breaking a 4-mm thick pane with a hammer. In three repeat experiments, an average of about 50 fragments were transferred to a retentive sweater, and 20 fragments were transferred to jeans described as "retentive." Further experiments involved breaking the window followed by firm contact with a second individual. In seven experiments 18 to 37 fragments remained on the sweater and 9 to 19 fragments remained on the jeans. The recipient individual wore a retentive sweater and fairly retentive cord pants. These items were worn for either 5, 30, or 60 minutes before being checked for glass. For the seven experiments, two to six fragments were recovered from the sweater, no glass was recovered (five times), and one and three fragments were recovered from the pants. Two experiments with the donor and recipient wearing tracksuit tops with low retention properties resulted in four and two fragments remaining on the donor item and no glass on the recipient item.

Allen et al.[128] also investigated secondary transfer. In their experiments, they studied the possible transfer between a person who had broken a window and a second person riding in the same car. Out of the 15 experiments involving the breakage of a windowpane with a crowbar, only one fragment was found on the second person. Whereas, after the 20-minute ride, typically 10 to 30 fragments were found (in one case 130 fragments were found) on the breaker. No more than three particles in total were transferred onto the car seat.

Allen et al.[129] also studied the transfer of glass fragments from the surface of an item to the person carrying it. Twelve experiments were performed where a box was placed 1 m behind a window, which was subsequently broken with a crowbar at a distance of 0.5 m. The box was then carried. In all experiments glass was recovered on the box, and between 1 and 22 fragments were found on the upper/lower garments of the person carrying it.

Tertiary transfer experiments were conducted by Holcroft and Shearer.[137] The experiments involved placing 33 and 25 glass fragments on a chair, followed by the recipient sitting for 5 minutes and walking for 5 minutes. In each case no glass was found on the sweater and one fragment was recovered from the jeans. A further experiment involved the breaker of the window sitting in a chair for 5 minutes. Then the recipient also sat in the chair for 5 minutes and walked around for a further 5 minutes before removal and searching of the outer clothing. In three experiments the number of fragments remaining on the sweater and jeans of the window breaker ranged from 11 to 28 and 8 to 20, respectively. No glass was found on the sweater of the recipient, and in each case one fragment was found on the jeans.

5.1.15 Transfer: what do we know?

The value of the published studies on the transfer of glass to garments is to establish an order of magnitude for the number of fragments (0 to 100) that can be transferred to a person breaking a window and to understand the complexity of transfer. This depends not only on distance, but also on the type of garment, the window, the force, or other unknown factors. They also show that secondary transfer of a few particles is possible, but not probable.

These studies have shown that on average more fragments are recovered on an upper garment that is at the height of the impact. Moreover, there could be a relationship between the number of fragments recovered on the ground and on the garments:[123] the number of particles smaller than 1 mm recovered on a person would correspond with the number of fragments found on the area of 300 cm² where the person stood. This raises an interesting, but impractical idea that it would be possible to predict from the crime scene the number of fragments transferred.

In summary, the conclusions on transfer are as follows.

1. It is possible to find glass on a person who has broken a window or who was standing nearby.
2. Transfer is a variable process. The variability is hard to assess because of the relatively small number of experiments performed and the apparent high variability of the phenomenon itself.
3. The greatest distance at which particles can be transferred is about 3 or 4 m.
4. The order of magnitude of the number of fragments transferred is 0 to 100; usually the size of the fragments is between 0.1 and 1 mm.
5. The number of particles projected backward diminishes with distance: every 0.20 m, the number decreases by half.
6. Experiments on the influence of the pane show that thick and/or wired glass produces more fragments.
7. About 50% of particles transferred by backscatter have an original surface.
8. Force (or pane resistance) may have an important influence on the number of particles projected backward.

All the research published has been performed on the transfer of window glass. However, transfer of other glass types are important, especially if the suspect has an alternative explanation for the presence of glass on his clothing. Therefore, it would also be useful to study transfer of glass other than from windows.

5.2 Persistence of glass on garments

Whereas early studies on persistence were scarce, there has been a considerable amount of research in several countries on the subject since the 1990s. The two pioneering publications that are presented later used artificial transfer methods. This has led to some uncertainty regarding the reliability of the findings.

The more recent studies of the persistence of glass fragments transferred glass when breaking a windowpane. This should remove this source of uncertainty.

The two earlier works demonstrate considerable foresight, however, in that they studied phenomena, the exact relevance of which took many years to become accepted.

5.2.1 Early studies

The work of Pounds (personal communication, 1977) consisted of placing a known number of glass fragments of different sizes on different types of garments — bulky woolen sweater, woolen sweater, Tweed jacket, sweater, and sports jacket. The size fractions were 0.1 to 0.5 mm, 0.5 to 1 mm, and particles bigger than 1 mm. The loss of these fragments was determined by examining the garments after a given time lapse. This study shows that glass fragments are lost rapidly, particularly when larger than 0.5 mm. For the bulky woolen sweater, only 8% of the fragments were recovered after 6 hours.

Brewster et al.[138] have studied the retention properties of two types of material: cotton (denim) and wool/acrylic (proportion 70:30). Glass particles were transferred using an air rifle and counted under a microscope after 30 minutes and 1, 2, 4, and 6 hours. The results show that the percentage of retained particles depends on the size of the fragments and that particles between 1 and 0.5 mm in size are retained for the longest time. It is surprising that fragments smaller than 0.5 mm are not retained longer. Plausible reasons for this result include that there could a critical size for the retention of fragments or that very small fragments are difficult to see under a microscope. This last hypothesis is supported by Pounds and Smalldon's research[28] on the searching of garments for glass.

Another surprising result is that glass is retained longer on denim than on wool/acrylic. This result could also be explained either by the fact that glass was examined with a microscope or by the method of transfer.

The disadvantage of these two studies is that glass was not actually transferred by breaking a window. Therefore, it was necessary to study the transfer and persistence of glass fragments on garments under more realistic breaking conditions.

5.2.2 Persistence of glass on clothing

The latest studies on the persistence of glass were done by breaking a window and observing how many fragments were retained after a certain period of time. Batten[139] was the first to study actual persistence. He arranged for eight windows (measuring approximately 1.50 × 0.60 m), four of which bore a plastic coating on one surface, to be broken during a laboratory renovation. The panes were broken by repeated blows from a hammer until most of the glass had been removed from the frame. After either 30 or 60 minutes of normal laboratory activity (not involving glass), hair combings and outer clothing were collected from each subject and searched. Glass was recovered from the hair combings of seven out of the eight subjects and from all items of outer clothing, with the numbers ranging widely from 2 up to about 100 fragments. Comparative results will be shown later.

Hicks et al.,[133] Hoefler et al.,[113] and Cox et al.[140-142] used the same clothing and windowpanes as described previously in the section under transfer experiments; the respective distances at which the windows were broken were 0.60 m and between 0.60 and 0.80 m. The first two studies consisted of breaking a new window for each time interval (Hicks et al., 30 , 60, 120, 240, and 480 minutes; and Hoefler et al., 15, 30, 60, 90, 120, and 180 minutes) during normal activity (Hicks et al.) or periods of walking (15 minutes) and resting (10 minutes). The number of fragments present after time had elapsed was counted. Cox et al. transferred glass by breaking a window; however, instead of going to their normal activities, the strikers entered a box in which they performed an activity (three activities were tested: 6 minutes of running, 20 minutes of walking, and 30 minutes of sitting). Five periods of the same activity were performed in five wooden boxes. The boxes consisted of four sides and a bottom, so it was possible to collect all glass that had fallen from the garments during activity and to know the number of fragments that had actually been transferred by addition of the collected fractions and glass remaining at the end of the experiment. This procedure also allowed an estimate of the influence of the degree of activity. In all three studies (in the Cox et al. experiments these were the final fragments remaining), the fragments were recovered by shaking and counting under a microscope.

The results are summarized in Table 5.4. All three studies show that loss appears to be a two-stage process consisting of a rapid initial loss and a slower process that can extend over a longer period of time (see Figure 5.2). Most of the fragments were lost during the first 30 or 60 minutes irrespective of activity, and thereafter the rate of loss seemed to stabilize. Cox et al. showed that the percentage of loss did not depend on the number of particles transferred.

The size of the fragments recovered depended on the time elapsed: the longer the elapsed time between search and breaking, the larger the proportion of small (0.2 to 0.5 mm) fragments. Cox et al. noted that during the initial rapid loss of fragments the influence of size was more important than in the second loss process, where the proportion of sizes appeared to remain constant. No clear relationship between size and activity could be established, as different activities implied different time lapses. However, it seems that the more vigorous the activity during the first 60 minutes, the smaller the size of the recovered fragments.

Hicks et al. showed that the garments used had a bearing on the number and size of the particles retained: the bulky sweater retained bigger and more numerous fragments than the cotton tracksuit (see Figures 5.2 and 5.3). In Cox et al. no difference between clothing was reported. However, this may have been due to the similar nature of the garments used.

Cox et al. have also investigated the relationship between shape and retention of fragments using the same classification as Underhill.[135] The authors concluded that "The persistence of glass fragments on the clothing studied was similar for all three fragments shapes, irrespective of activity or clothing type."

As a practical conclusion, it has to be noted that even 8 hours after the breakage it is possible to find as many as seven glass fragments on clothing.[133]

Table 5.4 Number of Fragments Recovered on Garments

Distance	Research	Clothing	Repeats	30 minutes Mean	30 minutes Range	60/100 minutes Mean	60/100 minutes Range	120/150 minutes Mean	120/150 minutes Range
0. 50 m	Cox et al.[a]	Shellsuit top (L)	4	11	1–28	7	1–12	9	1–14
	Running = 30 minutes	Shellsuit bottom (L)		8	5–13	2	1–3	4	1–7
	Walking = 100 minutes	Tracksuit top (M)	4	17	6–29	23	8–39	12	0–26
	Sitting = 150 minutes	Tracksuit bottom (M)		31	6–74	13	7–17	8	2–11
		Sweater (M)	4	33	4–51	11	8–13	26	10–42
		Jeans (M)		17	6–29	9	1–9	18	8–25
Estimated 0.50–0.70 m	Batten[b]	Top (L)	0	4	—	2	—	—	—
		Bottom (L)		6	—	3	—	—	—
		Top (L)	0	10	—	2	—	—	—
		Bottom (L)		6	—	2	—	—	—
		Top (H)	0	>100	—	31	—	—	—
		Bottom (H)		14	—	35	—	—	—
0. 60 m	Hicks et al.	Tracksuit top (M)	6	20	12–32	13	5–30	7	2–16
		Tracksuit bottom (M)	6	11	4–29	6	2–10	5	1–8
Estimated 0.50–0.70 m	Hoefler et al.[c]	Sweatshirt (L)	20/10/5	4	0–9	2	1–5	0	0–1
		Jeans (L)	20/10/5	6	3–10	5	2–11	3	2–4
0. 80 m	Hicks et al.	Pullover (H)	6	10	4–17	10	5–21	8	4–10
		Jeans (L)	6	12	5–25	13	4–28	6	2–9

[a] Cox et al. used time intervals of 30, 100, and 150 minutes.

[b] For each of the eight experiments, a new type of clothing was used. In order to compare times, clothing with the same retention are presented. These clothes are the following: (30 minutes) nylon kagoul, smooth hopsack weave, cotton T-shirt, jeans, woolen sweater, thick cord; (60 minutes) acrylic sweater, smooth weave, acrylic sweater, jeans, woolen sweater, thick cord.

[c] In this research the number of repeats depends on the time interval: for time intervals of 15 and 30 minutes, there were 20 repeats; 10 repeats for 60 minutes; and 5 repeats for the remaining intervals.

Note: The retention properties are indicated as L (low), M (medium), and H (high).

5.2.3 Persistence of glass on shoes

Pounds (personal communication, 1977) reported the number of fragments transferred to the soles of shoes. Broken window glass was placed onto a linoleum-covered floor, and the particles were transferred to the soles by treading on the floor. It was found that fragments on soles are lost very rapidly. In the case of rubber shoes, only 5% of the particles were retained

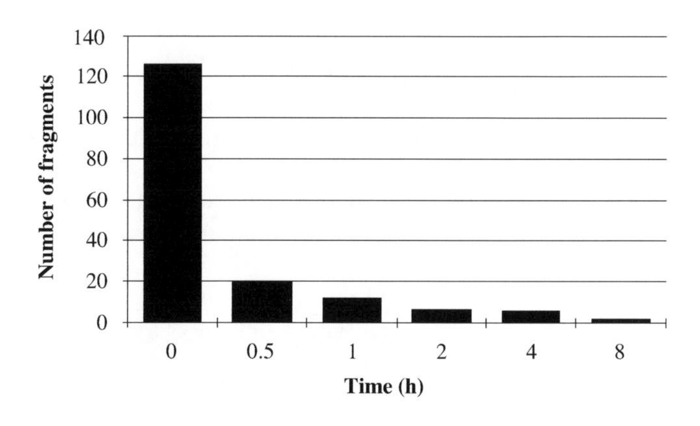

Figure 5.2 Persistence of glass fragments on the breaker's cotton sweater. A person standing at 50 cm broke the pane with a hammer.

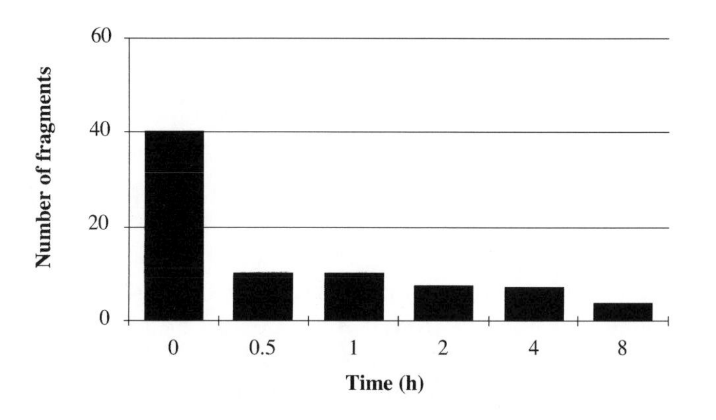

Figure 5.3 Persistence of glass fragments on the mock accomplice's woolen pullover. The pane was broken with a hammer with the mock accomplice standing at 80 cm.

after 30 minutes; however, deeply embedded fragments may stay for a considerable time. The author mentions that often the only way to recover such fragments is probe searching with a dissecting needle. When searching a leather sole visually with a microscope after 4 minutes, 12% of the particles were recovered, whereas 45% were recovered with a probe. For rubber soles, the respective numbers are 7 and 12%. Unfortunately, there is no mention of the total number of fragments recovered.

Holcroft and Shearer's[137] experiments also involved transferring glass to shoes by walking through broken glass. The individual walked for a further 5 minutes before the training shoes were packaged. This experiment was performed four times; the total number of glass fragments recovered in each experiment was four, nine, eight, and seven, respectively.

5.2.4 Persistence of glass in hair

Batten[139] studied the persistence of glass in hair. Eight windows were broken with a hammer, and glass was searched for in the hair of the breakers (or bystanders) after 30 and 60 minutes. It is not known if the windows were broken by eight individuals or fewer. The results are shown in Table 5.5.

Table 5.5 Persistence of Fragments in Hair

Experiment	Number of fragments recovered in hair after 30 minutes	Number of fragments recovered in hair after 60 minutes
1 and 5	9	0
2 and 6	22	3
3 and 7	3	4
4 and 8	3	2

As in other experiments, there is some variation. However, the study shows that a considerable amount of glass may be recovered from the hair of a person who has broken a window.

5.3 Main results of the studies

At the beginning of this chapter, we cited ten factors that could affect transfer and persistence studies.

The size of the window. Results from several studies suggest that it is the area of damage to a window, rather than the size of the window itself, which relates to the number of fragments produced by backward fragmentation and may subsequently be retained on clothing.

The type and thickness of the glass. Limited information from the Locke studies suggests that the harder the glass is to break, the more fragments are produced when it does break.

How the window was broken. More fragments seem to be transferred to the clothing of the perpetrator when a window is broken by multiple blows than when it is broken by a single blow. It also appears that a rapid break with a heavy weight produces fewer fragments than a slower break with a lighter weight.

The position of the offender relative to the window. If a person is more than 1 m from a breaking window, then it is likely that very few fragments will be transferred to and retained on the clothing. If a person is close to a breaking window, and the point of impact is above waist level, it is likely that considerably more fragments will be retained on the upper, rather than the lower clothing.

Whether or not entry was gained to the premises. This parameter does not seem to have a great influence on the number of fragments transferred.

The nature of the clothing worn by the suspect. This has a large effect on the number of fragments that would be recovered. Results of the studies suggest that there are several interrelated factors involved, including the coarseness of the fabric and the construction of the garment.

The activities of the suspect between the times of the incident and the arrest. There is very little reliable information on this topic, but Cox et al. suggest that the more vigorous the activity, the more rapid the rate of loss will be.

The length of time between the arrest and the clothing being taken. The persistence experiments show that the longer the time interval, the less glass will be remaining.

The way in which the clothing was obtained from the suspect. Various studies (for example, Hoefler et al. or Allen et al.) demonstrate that much glass can be lost when clothing is removed.

The weather at the time of the incident. Damp or wet clothing would appear to be more retentive than dry clothing.[126-129]

Recently, there has been much research on transfer and persistence. It is true that there are gaps in our knowledge, but it is also true that the studies described previously have provided us with much valuable information. When they are considered collectively, they suggest the following general principles.

1. Even when experimental conditions are carefully controlled, the number of fragments produced when a window is broken can vary considerably.
2. At least a third of all casework-sized fragments produced by backward fragmentation have original surfaces.
3. In most experiments described in the studies, glass was found on the clothing of an individual who was close to a breaking window.
4. The main factors affecting the number of glass fragments found on an individual exposed to a breaking glass window are the distance of the individual from the window and the nature of the clothing worn.
5. It is likely that more fragments will be found on the footwear or socks than on the pants of a person exposed to a breaking window when that person was standing more than 1 m from the window.

6. Only a very small proportion of fragments on one item are likely to be transferred to a second item by contact between them.
7. Persistence appears to be a two-stage process. The initial loss in the first 30 or 60 minutes is very rapid, the second process is slower, and after 8 hours at least it is still possible to find glass on clothing.

5.4 Modeling glass transfer and making estimates

If addressing propositions at the activity level, Bayesian interpretations of glass evidence require an estimate of T_n, "the probability that n fragments of glass were recovered given that an unknown number were transferred from the crime scene and retained on the offender."

This section describes the use of simple modeling techniques as a method for consistent and objective evaluation of the transfer probabilities.[143] We also describe how specific uncertainties can be handled.

5.4.1 Graphical models

The modeling used in this section is described in two phases. The primary phase constructs a simple deterministic model of the transfer, persistence, and recovery processes. This model describes the factors thought to be involved, the parameters that characterize each factor, and the dependencies that exist between these parameters. This primary model does not allow for any uncertainty. The secondary phase uses this primary model to construct a formal statistical *graphical model* to describe the stochastic nature of the transfer process.

The idea to use a directed graph to represent a statistical model is not a new one,[144] but developments in the use of these ideas in Bayesian analysis of expert systems have only come about relatively recently (see Reference 145 for a comprehensive review). We are aware that Dr. Ian Evett is using these models in the area of fiber evidence (personal communication, 1995).

Construction of a graphical model can be divided into three distinct stages. The first *qualitative* stage considers only general relationships between the variables of interest, in terms of the *relevance* of one variable to another under specified circumstances or conditions.[145] This stage is equivalent to the aforementioned primary modeling phase and leads to a graphical representation (a graphical model) of conditional independence that is not restricted to a probabilistic interpretation. That is, the qualitative stage, through the use of a formal graphical model, describes the dependencies between the variables without making any attempt to describe the stochastic nature of the variables. For example, studies have shown that the distance of the breaker from the window influences the number of fragments that land on the breaker's clothing. Therefore, distance and the number of fragments that land on the suspect would be included in the graphical model. The second *quantitative* stage would model the dependency between these two variables. This *probabilistic* stage introduces the idea of a joint distribution defined on the variables in the model and relates the form of this distribution to the structure of the graph from the first stage. The final *quantitative* step requires the numerical specification of the necessary conditional probability distributions.[145]

Use of a graphical model is appealing in the modeling of a complex stochastic system because it allows the "experts" to concentrate on the structure of the problem before having to deal with the assessment of quantitative issues.

A graphical model consists of two major components, nodes (representing variables) and directed edges. A directed edge between two nodes, or variables, represents the direct influence of one variable on the other. To avoid inconsistencies, no sequence of directed edges that return to the starting node are allowed, i.e., a graphical model must be acyclic. Nodes are classified as either constant nodes or stochastic nodes. Constants are fixed by the design of the study and are always founder nodes (i.e., they do not have parents). Stochastic nodes are variables that are given a distribution and may be children or parents (or both).[146] In pictorial representations of the graphical model, constant nodes are depicted as rectangles and stochastic nodes are depicted as circles.

5.4.2 A graphical model for assessing transfer probabilities

The processes of transfer and persistence can be described easily, but are difficult to model physically. The breaker breaks a window either with some implement (a hammer or a rock) or by hand. Tiny fragments of glass may be transferred to the breaker's clothing. The number of fragments transferred depends on the distance of the breaker from the window (because of the backscatter effect demonstrated by Nelson and Revell[2]). The activity of the breaker, the retention properties of the breaker's clothing, and the time until the breaker's clothing is confiscated are some of the factors that determine how many fragments will fall off the breaker's clothing.

We have discussed the factors important to take into consideration in order to estimate T_n. As it is unclear how to model some of these factors, the proposed model considers only the major effects of the position, time, garment type, and the laboratory examination.

Figure 5.4 is a very simplistic graphical model that describes how distance, time, garment type, and the lab examination will affect the final number of fragments observed on the suspect. The model can be described thus: the number of fragments transferred to the breaker directly depends on the distance of the breaker from the window during the breaking process. The number of fragments that are still on the breaker's clothing at each successive hour up to time t depends on the number of fragments that were initially transferred to the clothing, the number of fragments lost in the previous hour, the time since the commission of the crime, and the glass retention properties of the breaker's clothing. At time t, some number of fragments have remained on the breaker's clothing, and how many of those are observed depends on how many are recovered in the laboratory.

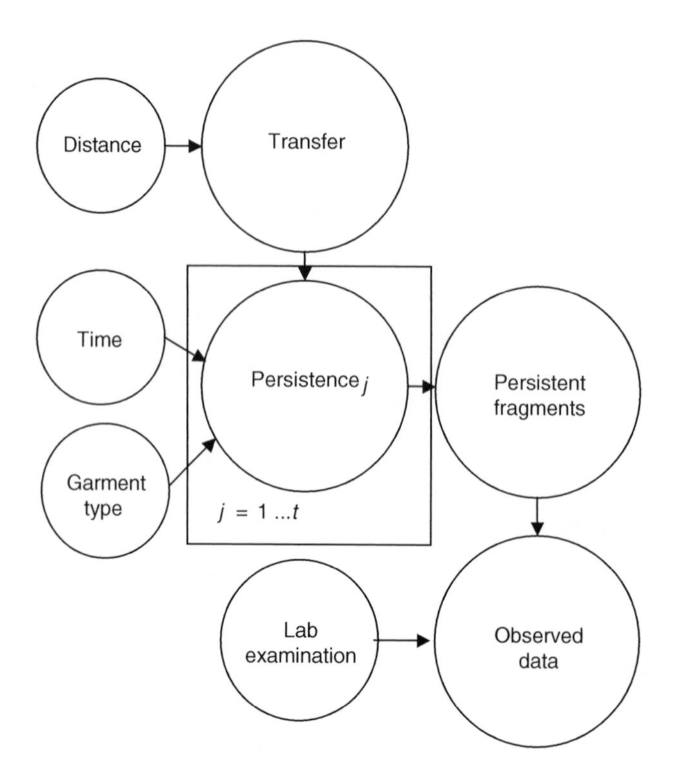

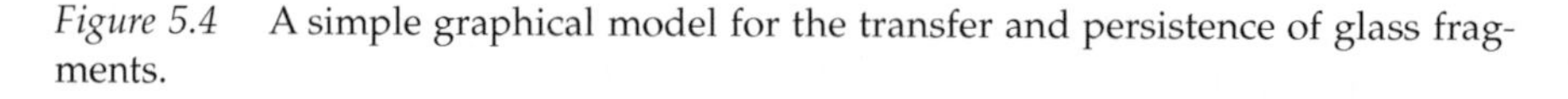

Figure 5.4 A simple graphical model for the transfer and persistence of glass fragments.

Each of these steps can be resolved into more detail to specify the full graphical model used in the transfer simulation program that provides the results presented here. The full graphical model is described in detail in Appendix A. The probabilistic modeling and quantitative assessment for the full model is described in Appendix B.

5.4.3 Results

The general probability distribution of T_n is analytically intractable. However, it is possible to approximate the true distribution by simulation methods. If the parameters that represent the process are provided, then it is possible to simulate values of n, the number of fragments recovered, by generating thousands of random variates from the model. If enough values of n are simulated, then a histogram of the results will provide a precise estimate of T_n. This simulation process can be thought of as generating thousands of cases where the crime details are approximately the same and observing the number of fragments recovered. Obviously, this is a computationally intensive process, and so to this end a small simulation program has been written in C++, with a Windows 95/98/NT® interface to display the empirical sampling distribution of n conditional on the initial information provided by the user. As well as allowing the user to specify the initial conditions, the program contains the full graphical model and allows the user to manipulate the dependencies between key processes.

Figure 5.5 shows the empirical distribution function of n, given the following information.

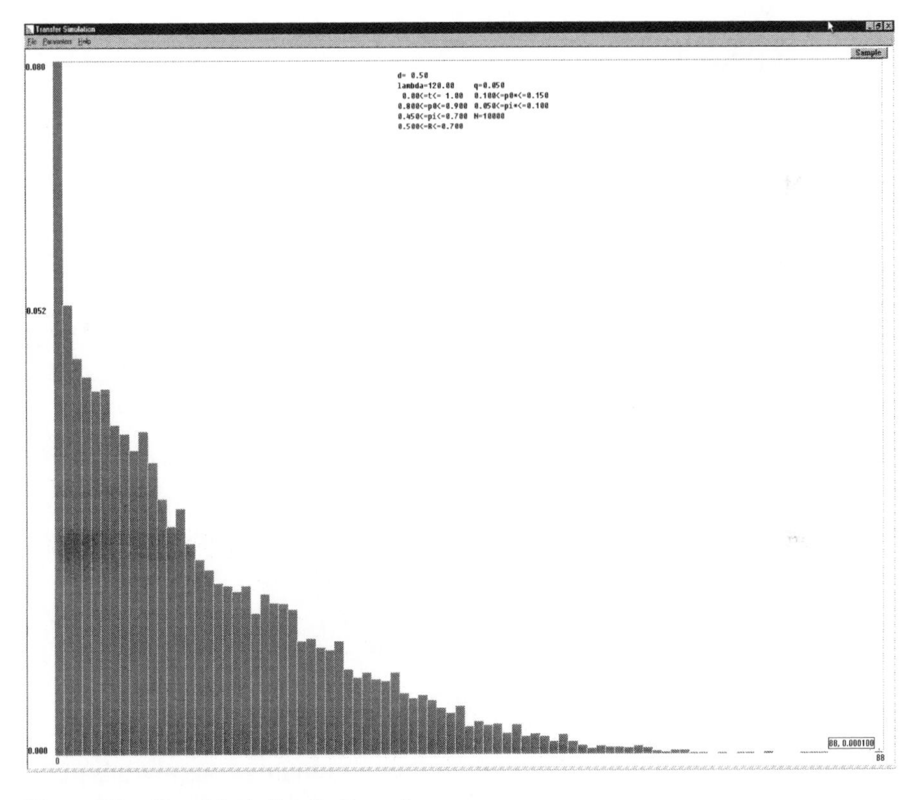

Figure 5.5 **Empirical distribution of** n.

The breaker was estimated to be 0.5 m from window. Please note that we do not take this value as known. The breaker is at a fixed, but unfortunately unknown, distance and we take 0.5 m as an estimate. The model incorporates uncertainty regarding this estimate into its procedure.

Given that the breaker was 0.5 m from the window when he/she broke it, on average 120 fragments would be transferred to the breaker's clothing. On average the breaker would be apprehended between 1 and 2 hours after breaking the glass.

In the first hour, the breaker would lose, on average, 80 to 90% of the glass transferred to his clothing and, on average, 45 to 70% of the glass remaining on his clothing in each successive hour until apprehension.

Figure 5.6 shows the model with the same assumptions, but does not allow the distance of the breaker from the window to affect the mean number of fragments transferred to the breaker.

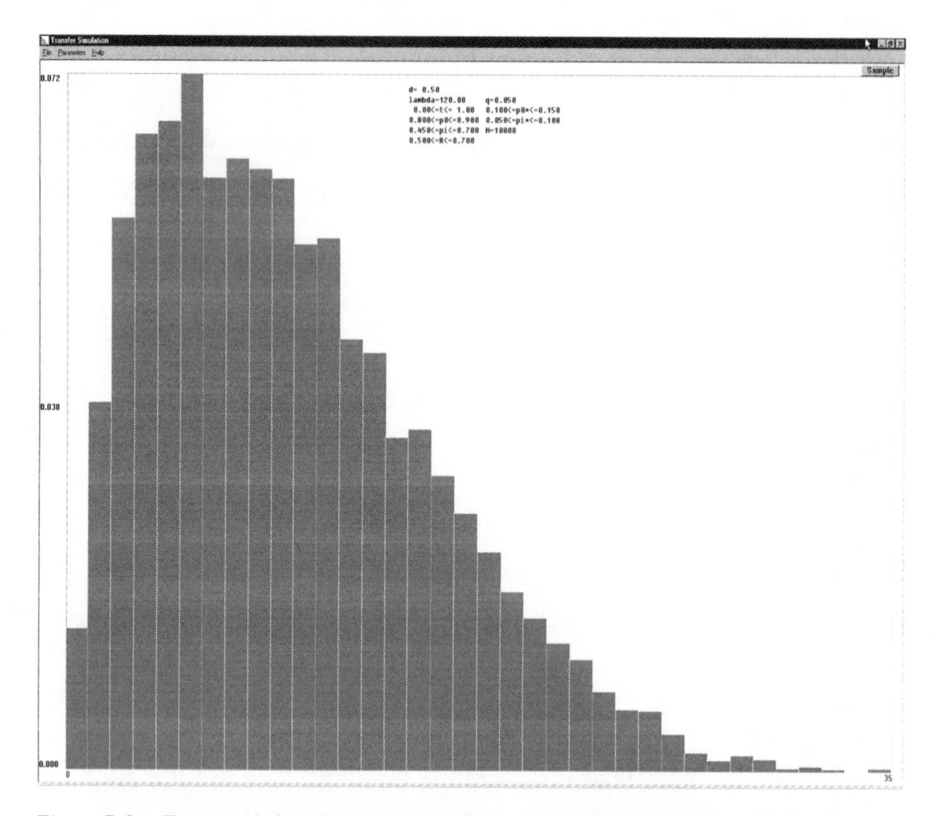

Figure 5.6 Empirical distribution of n without the distance effect.

Estimation of transfer probabilities is a formidable task for the glass examiner and, in fact, may be overwhelming. This is because of the inherent variability of the process, the multiple factors affecting transfer and retention, and the fact that many pieces of information may be lacking in the known case circumstances. However, it is seldom that nothing is known. For instance, the retentive properties of the clothing are usually estimable (since the clothing has been submitted). We argued earlier that even if there is no knowledge regarding the probable distance from the window of the breaker, people adopt only a few positions when breaking a window. In some cases there is no information regarding the time lapse between breakage, and packaging of the clothing. Even in this circumstance we can safely assume that it is almost impossible to package clothing less than 1 hour after the breakage, and for glass examiner purposes 30 hours approximates infinity. The graphical model gives a method by which many of these uncertainties may be accommodated.

For a very wide range of values in the parameters, the probability of recovering any given number of fragments is found to be low. However, it must be borne in mind that we are actually interested in the ratio of this probability to the probability of this number of fragments recovered from the clothing of a person unrelated to a crime. In many circumstances the probability calculated under the assumption of transfer, persistence, and recovery will, although low, be many times higher than the chance of this number of fragments recovered from the clothing of a person unrelated to a crime.

The effect of time is very marked: the longer the time until apprehension, the more fragments that are lost.

The effect on the distribution of the number of fragments recovered, by allowing the distance of the breaker from window to affect the average number of fragments transferred, is quite interesting. The distance factor is specified as a stochastic node in the full graphical model (see Appendix A). The assumed distribution of distance puts high probability on the true distance being close to the estimated distance. However, the true distance can be either less than or greater than the estimated distance. If the true distance is less than the estimated distance, the mean number of fragments that can be transferred to the breaker is increased. If the true distance is greater than the estimated distance, the mean number of fragments that can be transferred to the breaker is decreased. Because it is more likely (in this model) that the true distance will be less than or equal to the estimated distance, the overall mean number of fragments transferred in the simulations is higher than for an experiment where the true distance is fixed. This increase is represented by the longer tail of the empirical distribution function for the number of fragments recovered, given in Figure 5.5, as opposed to that in Figure 5.6.

The conclusions are robust to the naive distributional assumptions made in Appendix B. The graphical modeling technique offers consistency and reliability unparalleled by any other method. The answers it provides may be refined with more knowledge and better distributions, but that such naive assumptions can provide such a realistic distribution function (J. Buckleton and J.R. Almirall, personal communication, 1996) suggests that it is the dependencies that drive the answer and not just the distributional assumptions. A full Bayesian approach would take the graphical model as an informative prior distribution and modify it by casework data to obtain the posterior distribution.

The previous model was experimental, and with hindsight some of the modeling assumptions may have been naive. For instance, we draw attention to the assumed distribution regarding distance: $d_i \sim \text{Gamma}(d)$. The scientist estimates that the breaker was d meters from the window. Allowing the true value, d_i, to be a Gamma distributed random variable represents the fact that the true distance is more likely to be closer than further away. It seems likely that this assumption may need research.

The performance of the model, however, may be sufficiently startling that it may be plausible to utilize it in casework immediately.

The authors wish to encourage experimental work on these distributional assumptions.

5.4.4 Conclusions from the modeling experiment

We seek to make three main points. First, that forensic scientists should be aware of what question they are answering when they assess the value of the T_n terms in a Bayesian interpretation. We suggest that this question be "what is the probability of recovering n fragments from a suspect's clothing given that: (1) an unknown number of fragments were transferred to the suspect from the crime scene, (2) something is known about the retention properties of the suspect's clothing, (3) the distance of the breaker from the window has been estimated, and (4) the time between the commission of the crime and the arrest has been estimated."

The second point is that regardless of the method used to evaluate these probabilities, with the exception of T_0, the estimates should be low. That is, there is a very small chance of recovering any given number of glass fragments from a suspect even if he/she is caught immediately.

The third and most important point is that simple graphical models combined with extensive simulation effectively model the current state of knowledge on transfer and persistence problems. The graphical modeling technique offers consistency and reliability unparalleled by any other method and, thus, must be recommended as the method for estimating transfer probabilities.

The software used in this book is available for Windows 95/98 or higher only by e-mailing the authors.*

5.5 Appendix A — The full graphical model for assessing transfer probabilities

Figure 5.7 is a more complete graphical model of the transfer and persistence processes. This model can be interpreted as follows.

- The breaker is a fixed, but unknown distance, d_i, from the window. An estimate of this distance, d, is made by the forensic scientist.
- At this distance, on average λ_i fragments are transferred to the breaker's clothing during the breaking process. The average, λ_i, depends on an estimated average, λ from experimental work. An unknown number of fragments, x_0, are actually transferred.

* E-mail: curran@stats.waikato.ac.nz.

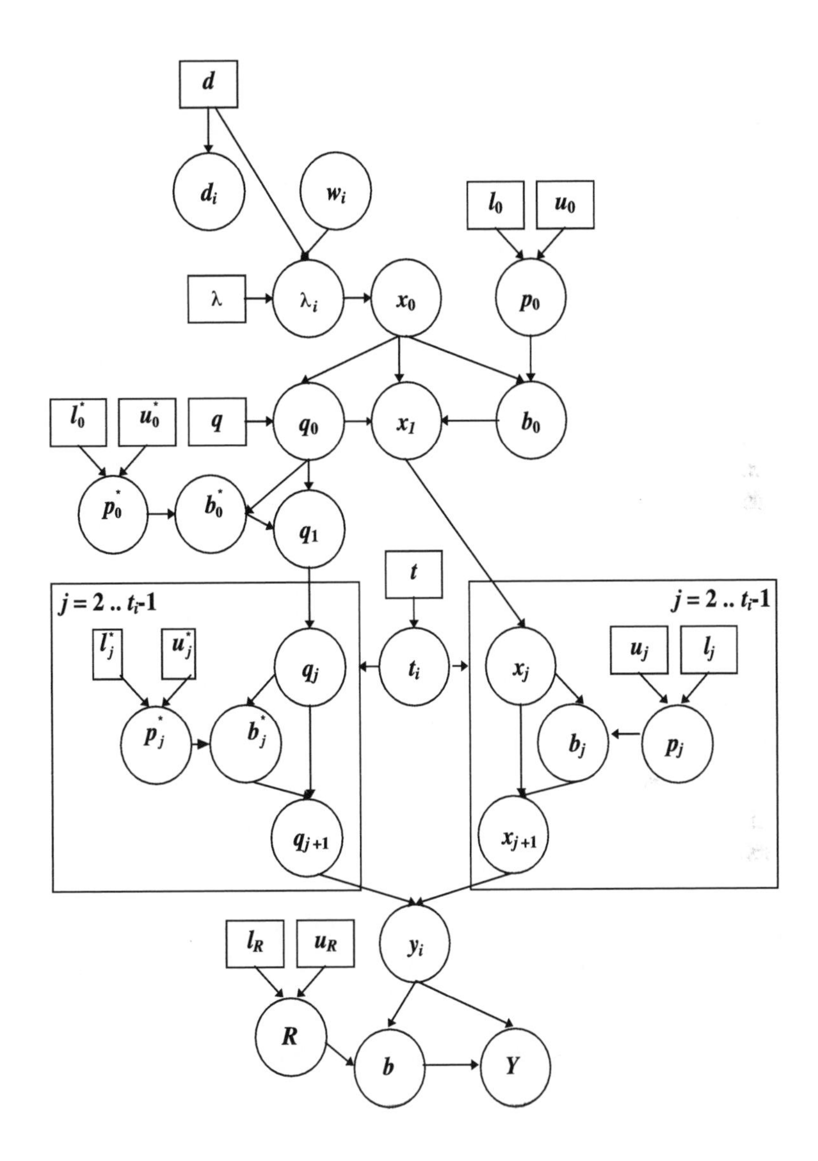

Figure 5.7 A formal graphical model of the transfer and persistence of glass fragments.

- Of the x_0, on average $100 \times q\%$ will become stuck in the pockets or cuffs or seams of the clothing or in the weave of the fabric. An unknown proportion, q_0, of fragments are actually in this category. The persistence of these fragments is modeled separately because they have a higher probability of remaining on the clothing.
- The breaker is not apprehended until an unknown number of hours, t, later. t is estimated by $\bar{t}$.

- During the first hour, on average $100 \times p_0\%$ of the x_0 and $100 \times p_0^*\%$ of the q_0 fragments initially transferred are lost, where p_0 and p_0^* are unknown, but lie somewhere on the intervals defined by $[l_0, u_0]$ and $[l_0^*, u_0^*]$, respectively. b_0 and b_0^* are the actual number of fragments lost.
- In each successive hour, on average $100 \times p_i\%$ of the x_j and $100 \times p_i^*\%$ of the q_j fragments remaining from the previous lost, are lost, where p_i and p_i^* are unknown, but lie somewhere on the intervals defined by $[l_i, u_i]$ and $[l_i^*, u_i^*]$, respectively. b_i and b_i^* are the actual number of fragments lost.
- At the end of t hours there are a total of y_i fragments remaining. On average, $100 \times R\%$ of these are recovered by the forensic scientist, where R is unknown, but lies somewhere on the interval defined by $[l_R, u_R]$. b is the actual number of fragments **not** recovered.
- Finally, Y fragments are observed.

5.6 *Appendix B — Probabilistic modeling and quantitative assessment*

As noted in Section 5.4, specification of a graphical model consists of three stages. Now that the first stage is complete, the probabilistic stage can be dealt with. The beauty of the graphical model comes from its conditional independence properties. It can be shown that the distribution of a child node depends only on the distribution of its parents.[145,146] This implies that if one distribution does not model a variable very well, then a better distribution may be substituted and the resulting changes will be automatically propagated through the affected parts of the model. In that vein the following distributions are proposed for each of the variables.

$d_i \sim$ Gamma(d): The scientist estimates that the breaker was d meters from the window. Allowing the true value, d_i, to be a Gamma distributed random variable represents the fact that the true distance is more likely to be closer than further away.

$w_i \sim$ Normal(1, 0.25); (l_i = ($l w_i \sim$ Normal(l, l/4)): The net result of letting λ_i be a normal random variable is a more spread out distribution for the number of fragments actually transferred.

$x_0 \sim$ Poisson ($e^{(1-d_i/d)} l_i$): The weight $e^{(1-d_i/d)}$ adjusts the mean number of fragments transferred on the basis of the true distance. If $d_i < d$, then the breaker is closer to the window than estimated, and, therefore, a higher number of fragments will be transferred on average. Similarly, if $d_i > d$, then the breaker is further from the window than estimated, and, thus, a smaller number of fragments will be transferred on average. This spread out Poisson distribution accurately reproduces the results given in Hicks et al.[133]

q_0 ~ **Binomial(x_0, q):** On average, $100 \times q\%$ of the fragments will get stuck in the pockets or cuffs or seams of the clothing or in the weave of the fabric.

$x_1 = \max(x_0 - q_0 - b_0, 0)$; b_0 ~ **Binomial(x_0, p_0)**; p_0 ~ **Uniform[l_0, u_0]:**

$q_1 = \max(q_0 - b_0^*, 0)$; b_0^* ~ **Binomial(q_0, p_0^*)**; p_0^* ~ **Uniform[l_0^*, u_0^*]:** Letting $p_0(p_i^*)$ be uniform represents the uncertainty over the number of fragments actually lost in the first hour and the retention properties of the garment. The number of fragments lost in the first hour is significantly higher than those lost in successive hours. Hicks et al.[133] show that the larger fragments are lost very quickly.

t ~ **Nbin(n, r):** The number of hours until apprehension, t, is modeled as a negative binomial random variable. The negative binomial distribution models the number of coin flips that are needed until n tails are observed, where the probability of a head is r. If $n = \bar{t} = l_t + u_t/2$, and $r = 0.5$, where l_t and u_t are estimated lower and upper bounds on the time, then the negative binomial provides a good estimate for the distribution of t, with a peak at $\bar{t}$ and a tail out to the right. This says that the time between commission of the crime and arrest is more likely to be shorter than estimated rather than longer.

$x_{j+1} = \max(x_j - b_j, 0)$; b_j ~ **Binomial(x_j, p_i)**; p_i ~ **Uniform[l_i, u_i]:**

$q_{j+1} = \max(q_j - b_j^*, 0)$; b_j^* ~ **Binomial(q_j, p_i^*)**; p_i^* ~ **Uniform[l_i^*, u_i^*]:** In each successive hour the suspect will lose, on average, $100 \times p_i\%$ of the fragments that remain on his clothing, where p_i is a uniform random variable that can take on values between l_i and u_i. The number of fragments actually lost in the first hour is b, where b is a binomial random variable with parameters x_i and p_i.

$y_i = q_t + x_t$; b ~ **Binomial(y_i, 1 –R)**; R ~ **Uniform[l_R, u_R]**; $Y = \max(y_i - b, 0)$**:** Finally, after t hours the suspect is apprehended; his clothing is confiscated and examined, and of the y_i fragments that remain on his clothing, b fragments are not found following the results of Pounds.[28]

chapter six

Statistical tools and software

The theory presented in the preceding chapters requires the examiner to make expert assessments of various probabilities and to evaluate the "relative rarity" of glass. In order to do this, a body of survey data is required. However, such collections are often available. Which collections are most suitable has been discussed in more detail in Chapter 4. We propose in this chapter to discuss what to do with the data; how to estimate the relative frequency of a given glass; and, in particular, how to construct histograms and density estimates. Several computer programs, which have been developed to assist the glass examiner when comparing measurements and assessing the value of glass in general, will also be discussed.

6.1 Data analysis

6.1.1 Histograms and lookup tables

We introduce a histogram discussion as a prelude to discussing density estimates. We take this long route because most people are unfamiliar with the concept of density estimation and find it a foreign idea. However, it is such a powerful tool in forensic interpretation that considerable effort is warranted.

A histogram is a statistical graph that displays frequency information. It is instructive to see how a histogram is constructed before discussing its relative merits.

6.1.1.1 Constructing a histogram

In the following method it is assumed that we have a set of N RI measurements, where N is a moderately large number ($N = 200$ is reasonable). We denote each measurement x_i for $i = 1, \ldots, N$.

Find the minimum RI,

$$x_{\min} = \min_i x_i$$

and the maximum RI,

$$x_{max} = \max_i x_i$$

Mark these values on your horizontal axis.
Divide the interval between the minimum and the maximum into a number of equal width class intervals or bins. The number of bins, k, is an arbitrary or ad hoc decision.
Construct a lookup table with the bins as the categories.
Place each observation into the appropriate interval in the lookup table
Count the number of observations in each class interval and draw a bar with height proportional to that count.

It should be obvious that this is a task better suited for a computer, especially for a large data set. The question of choosing the number of bins is a controversial one. In order to carry out this procedure correctly, we need to consider what properties we would like our histogram to have.

Desirable Properties of an RI Histogram
1. An accurate representation of the distribution of RIs
2. Robust to the addition of small amounts of new data
3. Sensitive to the addition of large amounts of new data
4. Sufficient resolution to provide accurate assessment of relative rarity

It should be apparent that the shape of a histogram is entirely dependent on the number of bins. For example, the 1994 New Zealand database of casework samples consists of 2656 measurements. From Figure 6.1 we can see that when we use ten bins we get the basic shape of the distribution of the data, but a fairly crude measure about the relative rarity of different RIs. That is, glass fragments are categorized as rare, common, or very common. While this categorization works, it is unlikely that any practicing caseworker would find it satisfactory. When the number of bins is increased to 50, the distribution becomes more discriminating. That is, it can distinguish between samples of glass with about a 0.07 difference in mean (that is equivalent to a difference of 0.00002 in RI). It is worthwhile noting that there seem to be two outlying samples. While it is necessary to describe this part of the data, we would really like to describe the bulk of the data better than the outlying points. In order to do this we need to use a specialized (nonparametric data driven) method for locating the upper and lower bounds that describe a large portion of the data. One way would be to construct a 95% confidence interval around the mean of the data.

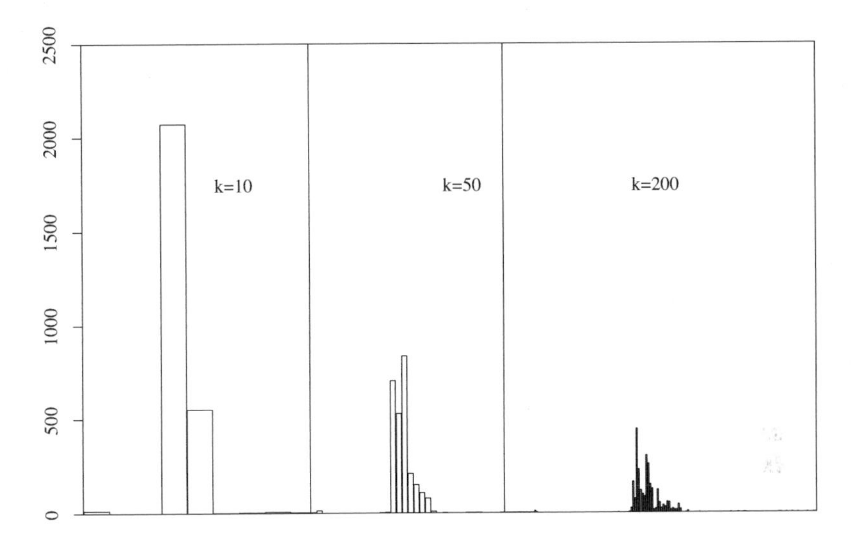

Figure 6.1 New Zealand casework data histogram with 10, 50, and 200 bins.

However, confidence intervals are poor tools for this purpose and depend on parametric assumptions that are unwise with complex distributions such as glass frequency distributions (they work best with simple distributions such as the normal distribution). A more robust choice is to find the whiskers of a box and whisker plot (more commonly known as a boxplot). The steps for constructing the whiskers for a box and whisker plot are the following.

1. Sort the data into ascending order, so that $x_{(i)}$ is the ith value in the sorted data set and $x_{(1)} \leq x_{(2)} \leq \ldots \leq x_{(n)}$.
2. Find the lower quartile (LO) and the upper quartile (UQ). These are the kth smallest and kth largest observations in the data set, where

$$k = \frac{\text{floor}\left(\frac{n}{2}\right) + 1}{2} \tag{6.1}$$

3. Find the interquartile range (IQR),

$$\text{IQR} = \text{UQ} - \text{LQ} \tag{6.2}$$

4. Find the approximate lower and upper bounds. These are given by

$$\text{LB} = \text{LQ} - 1.5 \times \text{IQR} \tag{6.3}$$

$$\text{UB} = \text{UQ} + 1.5 \times \text{IQR} \tag{6.4}$$

This may seem difficult, but it is best illustrated by an example. Our New Zealand data contains n = 2656 measurements. Therefore,

$$k = \frac{\text{floor}\left(\frac{2656}{2}\right) + 1}{2} = \frac{1328 + 1}{2} = 664.5 \tag{6.5}$$

The fact that k is fractional means we have to take the average of the 664th and 665th smallest values for the lower quartile and the average of the 664th and 665th largest values for the upper quartile:

$$\text{LQ} = \frac{x_{(664)} + x_{(665)}}{2} = \frac{1.51593 + 1.51593}{2} = 1.51593 \tag{6.6}$$

and

$$\begin{aligned}\text{UQ} &= \frac{x_{(2656-664)} + x_{(2656-665)}}{2} \\ &= \frac{x_{(1992)} + x_{(1991)}}{2} \\ &= \frac{1.51964 + 1.51964}{2} \\ &= 1.51964.\end{aligned} \tag{6.7}$$

Therefore, the interquartile range is IQR = UQ – LQ = 1.51964 – 1.51593 = 0.00371. Now, we calculate the approximate positions of the whiskers.

$$\text{LB} = \text{LQ} - 1.5 \times \text{IQR} = 1.51593 - 1.5 \times 0.00371 = 1.510365 \tag{6.8}$$

$$\text{UB} = \text{UQ} - 1.5 \times \text{IQR} = 1.51964 - 1.5 \times 0.00371 = 1.525205 \tag{6.9}$$

Next, we increase the lower bound until it falls on the first observation greater than LB. For our data set this value is 1.51086. Finally, we decrease the upper bound to the first observation less than UB. For our data set this is 1.52518. About 95% of our data lies between these two bounds. If we use 50 bins, there are approximately 10 bins that describe this 95%. That is, the bulk of cases will fall into one of ten categories. This is probably still too low for most practitioners. When we increase the number of bins to 200, this

Table 6.1 Suggested Number of Bins for New Zealand Casework Data

Method	Number of bins (k)
Sturges[147]	12
Dixon and Kronmal[149]	34
Scott[150]	64
Scott (corrected)[150]	130
Freedman and Diaconis[151]	68
Freedman and Diaconis[152]	138

number becomes 39. We have close to the discrimination we desire without too much loss of information.

As can be seen from the previous paragraphs, selecting the number of bins for a histogram is not an easy task. The examiner should not be drawn down the line of assuming that 0.0001 in RI units is anything more than some arbitrary choice based upon "human" convenience. There are a number of suggested automatic methods for choosing the number of bins (see Table 6.1). Sturges[147] suggested that $k = \text{floor}[1 + \log_2(N)]$, when N is a power of 2. Hoaglin et al.[148] note that this rule can be extended to N that are not a power of 2, but this is more for those who want an aesthetic picture rather than to convey information. Dixon and Kronmal[149] suggested that one choose $k = \text{ceiling}[10\log_{10}(N)]$ as an upper bound for the number of bins. Hoaglin et al.[148] note that this rule works moderately well for $20 \le N \le 300$. David Scott of Rice University in Texas, a well-known researcher in the area of density estimation, constructs a method for an optimal bin width, based on the assumption that the data is unimodal and symmetric.[150] If this is the case, he suggests a bin width of $h_N = 3.49sN^{-1/3}$, where s is the sample standard deviation of the data set. However, this method is unlikely to provide sufficient resolution if the data is strongly bimodal or multimodal. In such cases the data is likely to be oversmoothed. Scott[150] suggests that h_N be multiplied by a correction of about a half. Freedman and Diaconis[151] suggest a bin width based on the maximum difference between the histogram and the true distribution, giving

$$h_N = 1.66s\left(\frac{\ln N}{N}\right)^{1/3} \tag{6.10}$$

In a second publication, based on different criteria, Freedman and Diaconis[152] suggest

$$h_N = \frac{2IQR}{N^{1/3}} \tag{6.11}$$

Let us use our New Zealand data set to examine these formulae. Recall $N = 2656$.

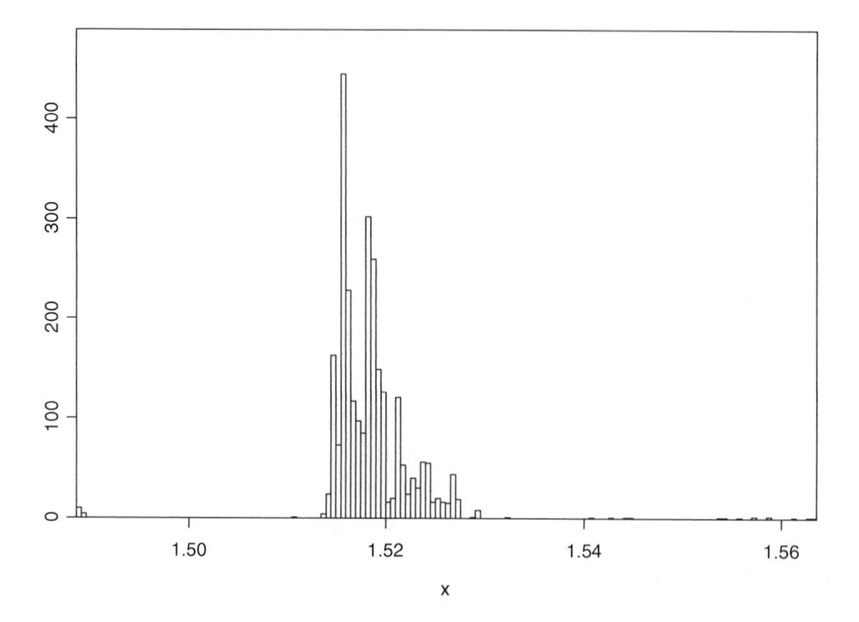

Figure 6.2 New Zealand casework data histogram with 135 bins.

It is clear from our previous investigation that the choice of 12 or even 34 bins is going to be too small. When we choose k to be 135, there are 26 bins for the bulk of the data.

From Figure 6.2 it seems that 135 bins provide a nice amount of discrimination — the bins have a width of 0.00055, i.e., they will give different values for samples with a difference in their means of 0.00055. This is about ten times the median standard deviation for recovered samples (perhaps a little large, but a nice compromise). The data are well described, i.e., they are neither too clumped together nor overdispersed.

The construction of a histogram in this way is a useful way to display the relative rarity of glass at different RIs. However, the estimation of a frequency for a particular type of glass is better performed by a "floating window."

6.1.2 Constructing a floating window

Most laboratories undertaking glass examination currently report a frequency for the glass found. This is actually quite a difficult topic, and given the marked preference of the authors for the continuous approach it will only be discussed briefly. In Chapter 2, Section 2.3 we described coincidence probabilities, which are the closest published formal definition of a frequency.

We mentioned earlier that given a precise definition of what is meant by frequency, it is easy to develop an algorithm that can estimate this probability. However, no completely satisfactory definition of "frequency" has ever been given.

In Chapter 2 we defined a frequency (coincidence probability) as an answer to the question: "What is the probability that a group of m fragments taken from the clothing of a person also taken at random from the population of persons unconnected with this crime would have a mean RI within a match window of the recovered mean?" Defining the match window is also difficult, and most attempts are based on applying the match criterion to m fragments from random clothing (or other population) and the n fragments from the control. Such a match window would be centered on the recovered mean and would have a width determined by the match criterion.

This is the standard application of the floating window. That is a window of width determined by the match criterion centered on the recovered mean. This differs from the standard application of a histogram, which shows "relative frequency" and gives an overall view of the data but does not make an attempt to define the vague term "frequency."

One algorithm is given in Evett and Lambert.[91] We give another algorithm in Appendix A.

6.1.3 Estimating low frequencies

There is at least one major flaw in the floating window approach. What happens when we encounter a sample of glass with a mean RI that falls either outside the range of RI histogram or into a window with zero count, i.e., a bin where no RIs of similar value have been seen before? It seems that we have something of a paradox. The glass sample we have recovered is very rare, but we cannot quantify this so that it will fit into our Bayesian framework, i.e., we would be dividing the numerator by zero. We could take some of the approaches used in the analysis of DNA data. For VNTR data, the band frequencies are (arbitrarily and incorrectly) obtained by dividing the allelic ladder up into bins so that there are at least five observations in each bin. This method is known as rebinning. When a histogram is constructed using unequal bin widths, the area of the bar associated with the bin represents the relative frequency of observations falling into that bin. This is in contrast to our previous histograms where the height of the bar alone gave the relative frequency. This, however, was really just a side effect of having all the bars of equal width, so the area of the bars in our previous histograms would give us the relative frequency. Rebinning would work and is certainly conservative, in that it would give much higher frequencies for many RIs. However, it is not an optimal approach, and there are better methods, as we will explain shortly. Another approach, used in the analysis of DNA, is to put the case sample into the database. That is, if we have N measurements in our database, and a case sample with two fragments with an RI that has not been seen before, we would estimate the frequency of the RI as

$$f = \frac{2}{N+2}$$

While this approach is at least robust and slightly easier to justify, it is still not optimal because it fails to recognize that RI measurements are continuous measurements. That is, there are no gaps between the possible RI values; any gaps observed are merely a function of the limits of accuracy in any measurement tool. The best way to display continuous data is to construct a density estimate.

6.1.4 Density estimates

In order to discuss density estimates we need some more terminology. Most of it will be familiar in some form or another, or at least logical. The reader may skip this section if he/she is familiar with the concept of continuous random variables and probability density functions.

6.1.4.1 Random variables and probability density functions

Definition

If X is a variable that records the result of a random experiment, then X is said to be a **random variable**.

Example. If X records the number of heads in ten tosses of a fair coin, then X is a random variable

Definition

If X is a random variable with a countable number of possible values or gaps between the possible values, then X is said to be a **discrete random variable**.

Example. In the previous example X can take on any value in the set $\{0, 1, 2, \ldots, 8, 9, 10\}$. In this example we can count all 11 possible outcomes. There are no outcomes in between any of those we have listed. Therefore, X is a discrete random variable.

Definition

If X is a discrete random variable, then the probability that X takes on a particular value, x, is given by $\Pr(X = x)$. $\Pr(X = x)$ is called the **probability function** of X.

Example. Assuming that each toss of the coin has no effect on the following coin tosses and that the only two outcomes on each toss are heads or tails, X is said to be a binomial random variable or have a binomial distribution, with parameters $n = 10$ and $p = 0.5$. The probability density function for a binomial random variable is given by

$$\Pr(X = x) = \binom{n}{x} p^x (1-p)^{n-x} \tag{6.12}$$

For example, if $n = 10$ and $p = 0.5$, then the probability of getting exactly five heads on ten throws is given by

$$\Pr(X = 5) = \binom{10}{5} 0.5^5 (1-0.5)^{0.5} \approx 0.246 \tag{6.13}$$

i.e., about 25%.

Note that $\Pr(X = x)$ is often written in shorthand as $\Pr(x)$.

Definition

If X is a random variable with an uncountable number of possible outcomes or there are no gaps between the possible outcomes, then X is said to be a **continuous random variable**.

Example. If X is the RI of a piece of glass, then X is a continuous random variable.

Definition

If X is a continuous random variable, and the probability that X lies between two points a and b is given by

$$\Pr(a \le X \le b) = \int_a^b f(t).dt \tag{6.14}$$

then $f(x)$ is said to be a **probability density function (pdf)** for X.

Note that it is important to note that for a continuous random variable with pdf $f(x)$, $f(x) \neq \Pr(X = x)$. In fact, for a continuous random variable, $\Pr(X = x) = 0$. This effectively says that if we could measure RI to infinite precision, the chance of observing two RIs that were exactly equal is zero.

Given that RI is a continuous random variable, we would like to find the pdf for RI. However, as with most physical models, the answer is not analytically tractable. What we must do instead is try to elicit the density function from our data. One way of thinking about this is to go back to our coin tossing model. We are going to toss a coin ten times and observe the number of heads. We would like to be able to specify the probability of observing any one of the possible outcomes. We have seen that we can work

Properties of Probability Density Functions

1. $0 \leq \Pr(a \leq X \leq b) \leq 1$
2. $\Pr(X \geq a) = \Pr(X > a)$ and $\Pr(X \leq b) = \Pr(X < b)$

   so

   $\Pr(a \leq X \leq b) = \Pr(a < X \leq b) = \Pr(a \leq X < b) = \Pr(a < X < b)^{*}$
3. $1 - \Pr(X > a) = \Pr(X \leq a)$
4. $\Pr(\phi) = \int_a^a f(t).dt = 0, \Pr(\Omega) = \int_{-\infty}^{\infty} f(t).dt = 1$

*This is not true for a probability function.

this out using the binomial distribution. But what would we do if we did not know about the binomial distribution? One way is to repeat the experiment many times and record the outcomes. Using the frequency of the outcomes, we could construct a histogram for the outcomes. If we could repeat this experiment often enough (in fact, forever), then eventually we would come up with a histogram that would have almost the same shape as the probability function for a binomial random variable. Because the histogram makes no assumptions about the true distribution of the outcomes, it is a **nonparametric density estimate** of the probability function.

Now, let us think about our RI distribution. If we could measure a large number of RIs and put them into a histogram, we would have a nonparametric density estimate of pdf for RI. However, as we put more and more data into the histogram, we would like the histogram to become more and more accurate. In order to do this we would need to increase the number of bins. That means that the bin widths would get smaller and smaller because the range of RI is between approximately 1.48 and 1.56. If this process could be repeated forever, then eventually the bin widths would be zero and we would be left with a nice smooth curve — i.e., the limiting histogram is the probability density function.

This is best illustrated with an example. Assume Z is a standard normal random variable. That is $Z \sim N(0,1)$, then the pdf for Z is

$$f(z) \sim \frac{1}{\sqrt{2\pi}} \exp\left(-\frac{1}{2}z^2\right) \tag{6.15}$$

Figure 6.3 demonstrates our point. As we take bigger and bigger samples from the normal distribution and increase the number of bins, the resulting histogram becomes closer and closer to the probability density function for the normal distribution.

So it would seem we have a method for finding the pdf for RIs. However, it is obvious that we cannot collect an infinite amount of glass and determine

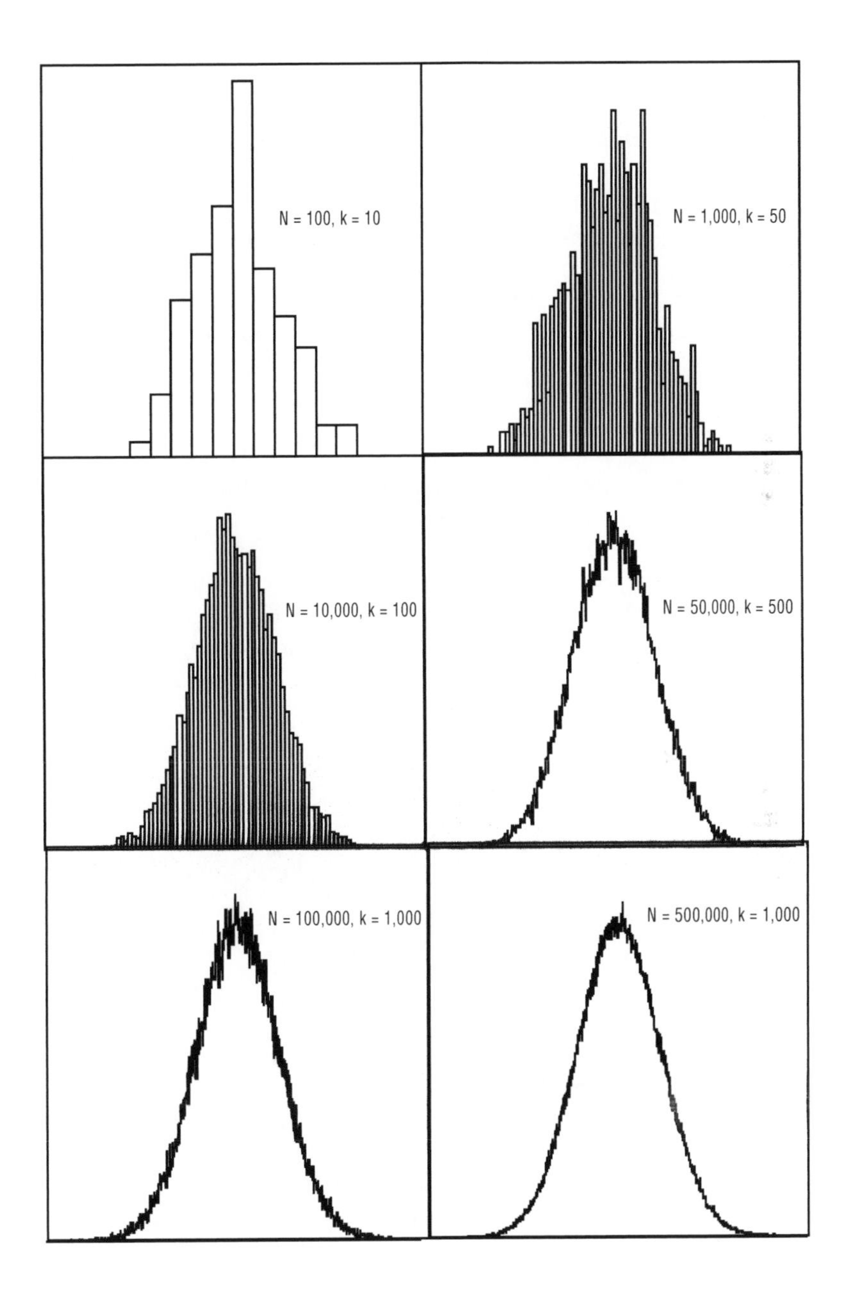

Figure 6.3 The limiting histogram of samples from a normal distribution. N is the sample size and k is the number of class divisions.

its RI. Currently, an experienced operator with a GRIM II can measure, at most, about 50 fragments in an 8-hour period. What we need is some way of smoothing our histogram or getting a smoothed histogram without needing all that data.

In order to do this we must relax one of our assumptions — the assumption about knowing nothing under the underlying distribution of the data. We know from the exploration of our case data that there are a number of modes (or bumps) in the data set and that the data is fairly symmetric around a couple of large central modes. We might think then that a localized assumption of normality is okay. That is, within small "windows" the RI of a piece of glass is a normally distributed random variable. So finally we get into density estimation.

6.1.5 Kernel density estimators

A traditional kernel density estimate works at smoothing the data by replacing each data point with a probability density function (typically a normal), or kernel, centered at the data point. The resulting density estimate at any point, x, is the mean of the values of the individual densities for each datum at that point. The smoothness of the resulting density estimate, $\hat{f}(x)$, is controlled by a tuning parameter or "window" or "bandwidth," h. $\hat{f}(x)$ is defined as

$$\hat{f}(x) = \frac{1}{Nh}\sum_{i=1}^{N} K\left(\frac{x - x_i}{h}\right) \tag{6.16}$$

where h is the window width, and the smooth kernel function,

$$K(x) = \frac{1}{\sqrt{2\pi}} e^{-\frac{x^2}{2}}$$

is the probability density function for a standard normal variable.

The astute reader should be asking at this point, "Where does h come from? Isn't choosing h just like choosing the number of bins for a histogram?" The simple answer, of course, is yes. However, a kernel density estimate with a well-chosen tuning parameter is often much more robust than a histogram. The reason for this is the nature of the kernel function. The kernel function essentially acts as a weighting function. Single datum with extreme values will have much less weight than a large group of data with similar values.

6.1.5.1 What is a good tuning parameter for a kernel density estimator?

Just as with choosing the number of bins for a histogram, there is no single or simple answer to this question. Scott[153] proves that if the data are truly

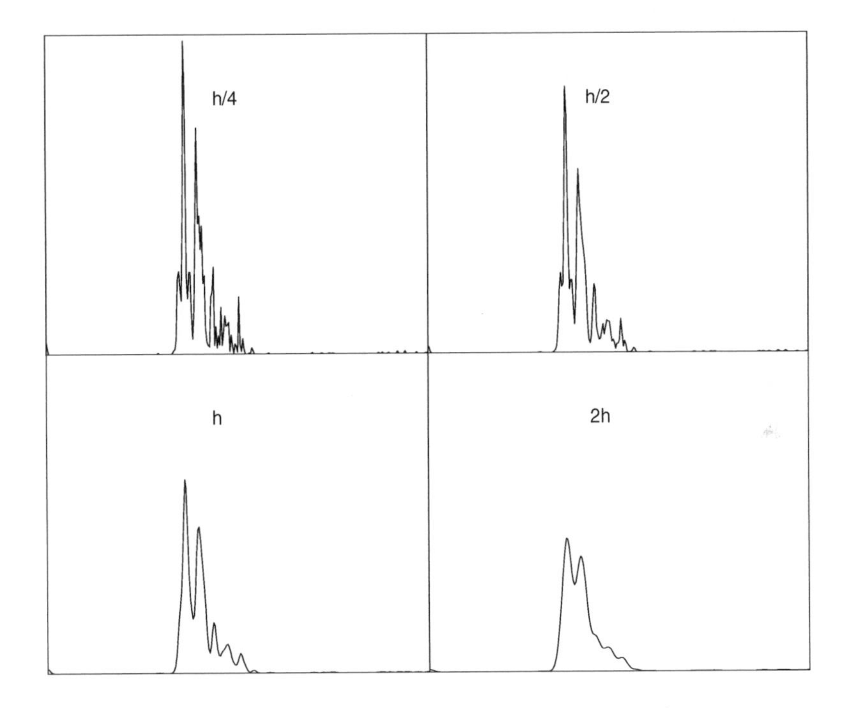

Figure 6.4 The effect of h on a kernel density estimate of the New Zealand casework data.

normally distributed, then choosing $h = 1.06 s_x N^{-1/5}$, where s_x is the sample standard deviation of the data, is optimal. However, we know that our data most definitely is not normal, and so the criteria for choosing h will be similar to that of choosing the number of bins.

Figure 6.4 shows the effect of changing the bandwidth on the kernel density estimate. If h gets too small, the resulting density estimate is jagged and has a lot of variation. When data is added to this density estimate it will register with stronger effect than if a larger h had been used. When h is too large, for example, when we choose a parameter that is twice as large as the one suggested by Scott,[153] too much detail is obscured. In the plot shown in Figure 6.4 the outlying data is not visible in the density estimate using $2h$. Figure 6.4 suggests that an h of somewhere between $^3/_4h$ and $1^1/_4h$ would produce a reasonable density.

6.2 Calculating densities by hand

In order to implement our recommended approach (the continuous Bayesian approach), two densities are required. We need to know (1) the probability of getting the analyzed characteristics of the unknown and the control knowing that they come from the same origin and (2) the probability of getting the analyzed characteristics of the unknown and the control knowing that they come from different origins. The density in the numerator (1) is a value from a *t*-distribution with either integral (equal variance *t*) or nonintegral (Welch's modification) degrees of freedom. Computer programs provide these values, but if such programs are not available, then they may be obtained "by hand" using a spreadsheet like Microsoft Excel which is intended to show readers how to obtain these values.

The second density that is required comes from survey data and appears in the denominator. It is also intended to demonstrate how that may be obtained by hand. Imagine that we proceed with an example where there is one recovered group and one control group. The analytical data are shown in Table 6.2.

Table 6.2 Summary Data for a Control and Recovered Group of Glass

	Control	Recovered
N	6	4
Sample mean	1.51804	1.51804
Sample SD	0.000055	0.000149

The mean of the recovered group is 1.51804. Therefore, in the denominator, we require the probability density of glass on clothing at 1.51804.

Table 6.3 Glass Frequency Data Excerpts from New Zealand Glass Surveys

RI	Window	Vehicle pre-1983	Vehicle 1984–1987	Vehicle 1988–1992	Container	Patterned	Flat
1.5176	2	17	9	5	0	1	1
1.5177	4	14	4	1	0	1	3
1.5178	3	11	5	2	2	0	3
1.5179	2	5	0	0	0	0	2
1.5180	3	11	2	0	0	1	2
1.5181	1	10	2	0	0	1	2
1.5182	2	19	11	2	0	1	1
1.5183	2	36	16	2	2	2	0
1.5184	1	24	16	4	2	0	1
Total samples	400	507	219	183	74	128	272

Earlier we discussed the values shown in Table 6.3 and stated that we would prefer glass on clothing surveys in most instances. We proceed here as if the reader does not have available a glass on clothing survey. Let us imagine that we can concentrate on the vehicle pre-1983 survey. We are only doing this to simplify the problem at this stage. We cannot think of many scientific reasons to concentrate on this survey other than it appears to be the most "common" at this stage. We require the probability density at

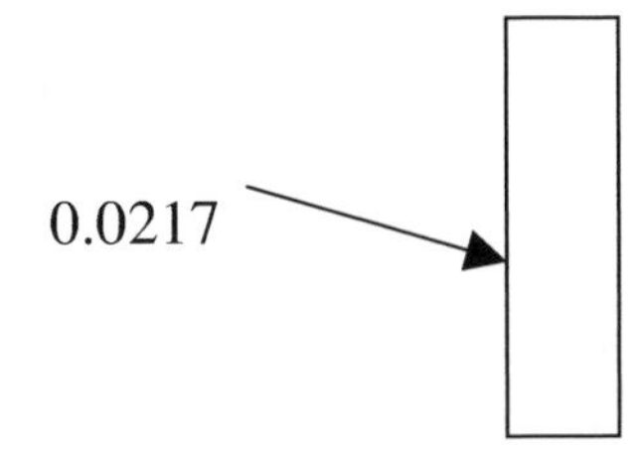

1.51804 for pre-1983 vehicles. There are 11 observations out of 507 in this histogram bar. Consider this histogram bar:

The area of the bar is the probability (0.0217). The width is 0.0001 RI units. The probability density is the height. This is 0.0217/0.0001 = 217.

A related approach might be to smooth the distribution first. This would "take out" local lumps and hollows. There are sophisticated and effective ways to do this, but a rough (but workable) method is simply to smooth over three (or five) histogram bars. In this case, smoothing over three bars we get a probability of 5 + 11 + 10/507 and a width of 0.0003 RI units, giving a density of 171.

In the absence of a clothing survey the analyst is faced with something of a dilemma. Which survey should be used? This has troubled many examiners, but is in fact more a "mental" problem than a real one. The analyst appears to have four choices:

1. Use a clothing survey.
2. Weight surveys of the above type according to subjective estimates of the likely source of glass on clothing (predominantly tableware).
3. Choose the most conservative option.
4. Determine the source (or sources) by elemental analysis or other method.

Such methods are not "state-of-the-art" statistics, but they may get laboratories "going" in a workable way.

The density for the numerator remains to be calculated by hand. Recall the statistics that were produced.

t-test: The test statistic is 0.05 for a *t*-distribution on 8 degrees of freedom

Welch's *t*: The test statistic is 0.04 for a *t*-distribution on 3.57 degrees of freedom.

Table 6.4 Probability Density from the *t*-Distribution

Test statistic	d.f. = 8	d.f. = 3.57
0.00	0.500	0.500
0.01	0.496	0.496
0.02	0.492	0.493
0.03	0.488	0.489
0.04	0.485	0.485
0.05	0.481	0.482
0.06	0.477	0.478
0.07	0.473	0.474
0.08	0.469	0.471
0.09	0.465	0.467
0.10	0.461	0.463

Note: The d.f. = 8 and d.f. = 3.57 columns give the probability mass to the right of the test statistic for 8 and 3.57 degrees of freedom, respectively.

Table 6.4 shows a section of a Microsoft Excel table produced using the TDIST function.

For the Welch distribution we require the density at a test statistic of 0.04. This is approximately (0.485–0.482)/(0.05–0.04), that is 0.003/0.01 = 0.3, which unfortunately is not on an RI scale. To rescale we observe that

$$V = \frac{(\bar{x} - \bar{y})}{\sqrt{\frac{s_x^2}{n} + \frac{s_y^2}{m}}} \tag{6.17}$$

therefore, a test statistic (V) of 1 is equivalent to a difference on the means of

$$\sqrt{\frac{s_x^2}{n} + \frac{s_y^2}{m}}$$

which in this case was 7.78×10^{-5}. To rescale we observe that the width of the "bar" is not 0.01, but $0.01 \times 7.78 \times 10^{-5} = 7.78 \times 10^{-7}$ and therefore the density is $0.003/7.78 \times 10^{-7} = 3856$. This is the density to compare with the "conservative" density of about 217 from the denominator.

6.3 Computer programs

No single comprehensive set of programs exists at this time that implements our recommended approach to casework. This is a serious flaw given the computational complexity of some of our recommended procedures. However, the components of a workable system exist in various places, and this section reviews these and suggests how they might be cobbled together.

6.3.1 The Fragment Data System (FDS)

This program was developed in the mid-1990s by Richard Pinchin and Steve Knight in order to compare RI measurements. The FDS program takes data either by manual input or directly from GRIM II. It can display this data in an elegant graphical form and perform objective grouping algorithms (Evett and Lambert, personal communication, 1991). The graphical output and ability to color and tag that data can be of great assistance to subjective grouping decisions. Once the fragments have been grouped, this program enables us to perform Student's *t*-test, the Welch test, and the continuous LR (using nonmatching glass from the LSH survey).[104]

This program can interact with CAGE (Computerized Assistance for Glass Evidence) for Windows in a limited fashion and thereby permit implementation of the continuous approach.

6.3.2 STAG

STAG (STatistical Analysis of Glass) was written by James Curran as part of his Ph.D. thesis in 1996 with suggestions and bug testing from John Buckleton, Kevan Walsh, Sally Coulson, and Tony Gummer from the Institute of Environmental Science Ltd., as well as Chris Triggs from the University of Auckland in New Zealand.

STAG revolves around sample data entered by hand into a spreadsheet-like interface. The user has the ability to designate samples as control and recovered samples, indicate where groupings might exist in samples, and store case notes on each sample. The data is entered in temperature form and converted into RI using a calibration line that each laboratory may set.

STAG incorporates many of the procedures and tests discussed in James Curran's thesis. In particular it is the only program to include the grouping algorithm of Triggs et al.[77,78] The manual or automatic grouping information is included in the subsequent analyses.

STAG has limited support for the Bayesian approach — it can produce continuous LRs based on the user's own glass database. STAG is available free from James Curran (e-mail: curran@stats.waikato.ac.nz).

6.3.3 Elementary

Elementary, written by James Curran, is a tool for the analysis of elemental concentration data (e.g., output from ICP-MS, ICP-AES, etc.). Elemental provides a simple interface for the implementation of Hotelling's T^2, as described in Reference 154 and the implementation of the continuous LR as described in Reference 155. It is a crude tool in that it requires the user to have (1) their own database and (2) input files organized by another program (such as a text editor or Microsoft Excel) in a format described in the help file. Elementary is available free from James Curran (e-mail: curran@stats.waikato.ac.nz). Both Elementary and STAG are Microsoft Windows programs written in C++ for the Win32 API. They have not been ported to other platforms, although the essential elements of both programs are fairly transportable.

6.3.4 CAGE

CAGE is written in CRYSTAL, a report management language. The underlying program uses a Rule Based Forward Chaining logic that works with the glass examiner to focus on relevant questions and to assist in producing an interpretation and associated paperwork. Written by Richard Pinchin and John Buckleton at the Home Office Central Research Establishment, Aldermaston, England (as it then was) in 1988, it is now considered by both authors to be in need of modernization despite being upgraded to the Windows 95 format and partially incorporating the continuous approach. Currently, this program has only been used in routine casework in New Zealand.

The program incorporated the approach of Evett and Buckleton[7] in a form intended for casework use. The formula is generalized and can handle most combinations of control and recovered groups. Data was taken from the McQuillan and Edgar survey.[156] After selecting an item, the user can choose several menus (see a description of the program or a table of terms), alter default data, and do a full case (with specific or average probabilities) or a precase analysis. CAGE assists the user when assessing transfer probabilities either by showing a summary of research or by giving the transfer probabilities assessed by other experts in similar cases (the similarity of the case is based on the type of control, the distance between offender and window, time elapsed, retention properties, and the type of glass). This program also offers the glass examiner very useful "help pages."

In the late 1990s CAGE was still relevant in that it produced (in symbolic form) the equations for any combination of controls and recovereds and attempted to explain each term. In addition, it performed the small services of providing the data in usable form from the ME survey and of performing the arithmetic. Minor upgrades, that could be done, are to introduce propositions management (CAGE enables us to address only two propositions); to put the menu preassessment first and have safeguards to warranty that transfer probabilities are first estimated and then fragments searched. Much research was performed in the late 1990s and so an upgrade would involve presenting these new results and integrating external modules such as the program Transfer. Accessibility on the World Wide Web could entice more people to use CAGE and perhaps would create a bigger database of RI data.

Currently, FDS and CAGE are available for purchase from the U.K. Forensic Science Service.

6.4 Summary

In this chapter we have shown how to use statistical tools in order to estimate $1/f$ or the continuous LR. We have presented computer programs and methods that enable us to compare measurements and to assist the forensic scientist when assessing the value of glass. The aim of our last chapter will be to present how we would report results in a glass case. We will see how it is possible to avoid fallacies, particularly what has been called the prosecutor's fallacy. We will continue with the presentation of four examples involving single and multiple controls and recovered, using elemental analysis.

6.5 Appendix A

If matching is performed using the Student's t-test, then a window is set as

$$\text{Recovered mean} \pm T_{crit} S_p \sqrt{\frac{1}{n} + \frac{1}{m}} \tag{6.18}$$

where T_{crit} is the critical value of the T statistic. If Welch's statistic is used then the window is given by

$$\text{Recovered mean} \pm V_{crit} \sqrt{\frac{s_x^2}{n} + \frac{s_y^2}{m}} \tag{6.19}$$

A frequency is then calculated either by counting the number of observations that fall into this window or by summing areas after smoothing.

There is a logical flaw in this approach that we are unable to resolve. The value of the control standard deviation appears in the definition of the width of the window. This is because it is integral to the definition of the match criteria. However, strictly speaking, an answer to the question — "what is the probability that a group of m fragments taken from the clothing of a person also taken at random from the population of persons unconnected with this crime would have a mean RI within a match window of the recovered mean?" — has nothing to do with the control glass sample. Hence, it is difficult to see why the control standard deviation should appear anywhere in the definition. This difficulty appears to be associated with the general difficulty in defining a frequency and is solved by the continuous approach.

chapter seven

Reporting glass evidence

Reporting is a source of constant difficulty in forensic science.[157] The challenge is to reduce the information created by the examination and typically held in a considerable case file to a few pages, or even a few sentences, of text. This, by its very nature, involves information loss. To make it slightly harder we ask that this be done without the use of formal probabilistic symbols and preferably in plain language. Anyone who for a moment thinks that this task in any way is easy should perform the following experiment.

> Get a colleague to choose a statement on a topic that you are not particularly familiar with. Sit in one seat for an hour listening to a boring radio program then ask your colleague to read the statement to you once. Ask your colleague to put the inflections and pauses in the wrong places because this is what will assuredly happen when a statement is read to the court. Challenge yourself to see what level of understanding and recall you can achieve.

The aim of this discussion is to attempt to identify those aspects of glass evidence most pertinent to the juries' understanding of the evidence and to preserve those aspects in the report. In the ensuing discussion we follow strongly the thinking of Evett.[158]

> *Interpretation of evidence takes place within a framework of circumstances.*

Because the scientist's interpretation depends upon the circumstances, it would seem reasonable for these to be made explicit in the statements.

> *To interpret evidence it is necessary to consider at least two propositions.*

We hold this to be a basic tenet of interpretation; therefore, we suggest that these two propositions should be explained in the statement.

> *It is necessary for the scientist to consider the probability of the evidence given each of these propositions.*

The results of the consideration may be reported with or without numbers or with both numbers and a verbal explanation. If numbers are used, we recommend something along the lines:

> *"This evidence is approximately (or at least depending upon approach) x times more likely if the clothing was close to the window at the pharmacy when it broke than if it has never been near this window."*

The same statement can be made without the use of numbers. Various scales have been offered, and we reproduce one here in Table 7.1 (from Reference 158).

Table 7.1 A Verbal Scale for Reporting LRs

LR	Verbal equivalent
1 to 10	Limited support
10 to 100	Moderate support
100 to 1000	Strong support
1000 upwards	Very strong support

Therefore, an interpretation giving an LR of approximately 500 might be reported as

> *In my opinion this evidence strongly supports the proposition that this clothing was close to the window at the pharmacy when it broke.*

This scale is symmetric. That is that LRs less than 1 should be reported as supporting the alternative hypothesis. This scale is not confined to glass evidence alone and is, in fact, perfectly general.

Table 7.2 illustrates how arbitrary these qualitative scales can be.

In addition to the arbitrary nature of these scales, confusion may arise over the verbal description of numbers. This is best illustrated by noting that the British interpret the word "billion" to mean 10^{12}, while the U.S. interprets the word as 10^{9}.

In general, we would advocate that when reporting an LR greater than one, the value of the LR should be included with the verbal equivalent. For example, the interpretation giving an LR of approximately 500 might be reported as

Table 7.2 Qualitative Scale for Reporting the Value of the Support of the Evidence for C against

Verbal Equivalent	LR	LR Approx.
Slight increase in support	1 to 100.5	1 to 3.15
Increase in support	100.5 to 101.5	3.15 to 31.5
Great increase in support	101.5 to 102.5	31.5 to 315
Very great increase in support	102.5 upwards	315 upward

From Aitken, C. G. G., *Statistics and the Evaluation of evidence for Forensic Scientists*, John Wiley & Sons, New York, 1995, p. 52. With permission.

> *This evidence is approximately 500 times more likely if the clothing was close to the window at the pharmacy when it broke than if it has never been near this window. In my opinion this evidence strongly supports the proposition that this clothing was close to the window at the pharmacy when it broke.*

In order to have something concrete upon which to focus discussion, we return to the simplest type of case.

Example 7.1 One group, one control

The suspect was apprehended 30 minutes after the breakage. The breakage was believed to have been performed by hitting the window with a club. Subsequently, the offender entered the premises through the broken window. The offender is believed to have exited through a door at the rear of the premises. Items taken or ransacked were not in the broken glass zone. Only one perpetrator was suspected.

As a matter of discipline, it is desirable to consider certain aspects of the case before the search is performed. Specifically, it is beneficial to consider the probability of transfer prior to knowing how many fragments were potentially transferred. This should increase the objectivity in these assessments. Determining the transfer probabilities presents difficulties for most glass examiners. However, the experienced glass examiner probably knows more than she or he realizes, and some training and thought experiments, some of which were given in Chapter 5, quickly demonstrate that the examiner is in a better position to attempt this admittedly difficult task than the jury.

From the work that has been done on glass transfer and persistence (Chapter 5), it appears reasonable to model the probability distribution for the number of fragments recovered after a given time by the method of Curran et al.[143]

Figure 7.1 shows the probability that m fragments were recovered from the suspect given the following information:

- The breaker was estimated to be 0.5 m from window.

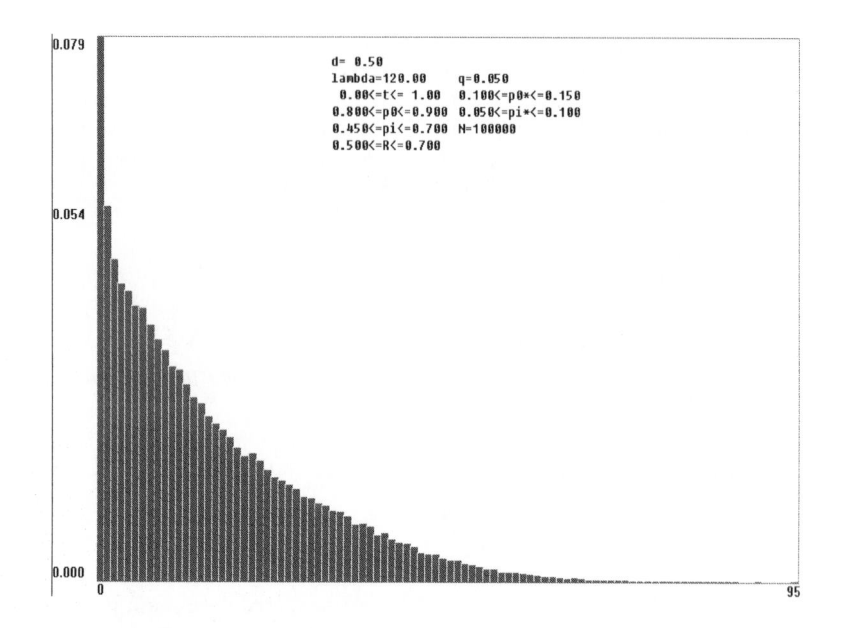

Figure 7.1 An empirical distribution function of *m*.

- Given that the breaker was 0.5 m from the window when he broke it, on average, 120 fragments would be transferred to the breaker's clothing.
- On average, the breaker would be apprehended between 0 and 1h after breaking the glass.
- In the first hour, the breaker would lose, on average, 80 to 90% of the glass transferred to his clothing and, on average, 45 to 70% of the glass remaining on his clothing in each successive hour until apprehension.

Using this information we estimate $T_0 = 0.079$ and $T_4 = 0.042$

A forensic examination is now performed. Four fragments were recovered from a full search of the clothing (T-shirt and jeans), including the pockets. Control and recovered measurements were as follows:

Recovered: 1.51820, 1.51812, 1.51798, and 1.51787
Control: 1.51805, 1.51809, 1.51794, 1.51801, 1.51805, and 1.51809

Summary statistics for the control and recovered measurements are shown in Table 7.3. A test using Welch's modification to the *t*-test appears

Table 7.3 Summary Statistics for a Control and Recovered Sample of Glass

	Control	Recovered
N	6	4
Sample mean	1.51804	1.51804
Sample SD	0.000055	0.000149

appropriate in view of the fact that the recovered glass is from clothing. Using this test, the glass matches at the 1% level.

t-test: The test statistic is 0.05. The 1% critical value for a t-distribution on 8 degrees of freedom is $t_8(0.005) = 3.35$.

Welch's t: The test statistic is 0.04. The 1% critical value for a t-distribution on 3.57 degrees of freedom is $t_{3.57}(0.005) = 5.01$.

The LR is given by

$$\frac{P(E \mid C, I)}{P(E \mid \bar{C}, I)} = T_0 + \frac{P_0 \cdot T_n}{P_1 \cdot S_n \cdot f} \tag{7.1}$$

From the LSH survey, $P_0 = 0.25$, $P_1 = 0.22$, and $S_4 = 0.02$. Determining f is not straightforward. Remember the difficulty we had in phrasing the question to form this probability (previously termed the coincidence probability). We seek to answer a question of the form: "If I find a group of six fragments on the clothing of a person unconnected with the smashing of the window in this case, what is the probability that it would be the same as the recovered glass?" The "same" is the difficult part (a precise solution exists in the continuous approach which was introduced in Chapter 3), but here we define it as glass within a match window of the recovered mean. The match window is defined by the nature of the comparison test, in this case Welch's t. This is actually quite complicated. Here, we will assume that a rough estimate is available as follows.

The Welch test defines a match window* by

$$\bar{y} \pm t_{\psi}(0.005)\sqrt{\frac{s_x^2}{n} + \frac{s_y^2}{m}} \tag{7.2}$$

where $t_{\psi}(0.005)$ is the (two-tailed) 1% critical value of the t-distribution with degrees of freedom given by

* The presence of the terms S_x and n in this equation is a result of the difficulty in defining the frequency. Strictly, we are considering where the glass did not come from the control; therefore, it is clear that these parameters should not appear in the formulation. The resulting formula given here is a best attempt to shoehorn the confidence approach into the correct thinking.

$$\psi = \frac{\left(\frac{s_x^2}{n} + \frac{s_y^2}{m}\right)^2}{\left(\frac{s_x^4}{n^2(n-1)} + \frac{s_y^4}{m^2(m-1)}\right)} \tag{7.3}$$

For this case, this formula defines a window centered on 1.51804 with lower and upper bounds given by subtracting and adding 0.00022, respectively. This interval contains all of the 1.5179, 1.5180, 1.5181, and 1.5182 histogram bars and 0.1 of the 1.5183 bar (containing the information from 1.51825 to 1.51835) and 0.3 of the 1.5177 bar. Using this approximate method, we obtain (eventually) $f \approx 0.04$.

Some (many) laboratories do not have a "clothing" survey, so we offer here an approximate and most probably conservative method. However, we strongly recommend the use of clothing surveys.

Suppose data of the type given in Table 7.4 have been collected. From this data we can obtain a set of frequency estimates by dividing each count by the number of samples and, in the case of the vehicles that change quickly with time, weighting according to modern vehicle numbers.

Table 7.4 Glass Frequency Data Excerpts from New Zealand Glass Surveys

RI	Window	Vehicle pre-1983	Vehicle 1984–1987	Vehicle 1988–1992	Container	Patterned	Flat
1.5176	2	17	9	5	0	1	1
1.5177	4	14	4	1	0	1	3
1.5178	3	11	5	2	2	0	3
1.5179	2	5	0	0	0	0	2
1.5180	3	11	2	0	0	1	2
1.5181	1	10	2	0	0	1	2
1.5182	2	19	11	2	0	1	1
1.5183	2	36	16	2	2	2	0
1.5184	1	24	16	4	2	0	1

Table 7.5 Summary the Glass Frequency Data Excerpts from New Zealand Glass Surveys

Type	Frequencies
Window	2.3%
Vehicle	7.2%
Container	2.4%
Patterned	1.9%
Flat	2.5%

None of these frequencies in Table 7.5 are exact, nor are they a substitute for a well-constructed clothing survey. However, they do give the range of most probable answers, and since we believe that most glass on clothing is container glass, it is not inconsistent with the 4% answer from the clothing survey. This type of survey, while inferior, is easily updated without a complete rework of the whole survey. For instance, in New Zealand the vehicle data is reworked every 3 years by adding a new block of data and reweighting. Window glass surveys separated by over 10 years in New Zealand were very similar and they are not updated as frequently.

Using the previous estimates, the LR becomes

$$\begin{aligned} \text{LR} &= 0.079 + \frac{0.25 \times 0.042}{0.22 \times 0.02 \times 0.04} \\ &= 59.73 \approx 60 \end{aligned} \tag{7.4}$$

We propose here to investigate this example further. First, we will try to verbalize this answer. This is an issue of great difficulty and importance. Second, we will investigate aspects of this approach such as sensitivity of this answer to some of the data estimates and to search procedures.

7.1 Verbalization of a likelihood ratio answer

Strictly, a phrasing of this answer could be the following: "This evidence is about 50 times more likely if the clothing was within two meters of the window when it broke than if it has never been near this window." In saying this, we have done several things. First, we have rounded down the answer, which reflects our own view of the precision (single digit) possible with this type of analysis. Second, we have avoided the fallacy of the transposed conditional. At this time we make the point that this evidence does not imply that "it is 50 times more likely that the clothing was within 2 meters of the window when it broke than if it has never been near this window." This is a transposed statement and is false. It is also noted that, following forensic practice, we have avoided the words guilt and innocence and have substituted hypotheses.

C: The clothing was within 2 m of the window when it broke.
$\bar{C}$: The clothing has never been near this window.

While we advocate placing before the court a numerical estimate of the value of the evidence as given previously, many examiners do not feel that this is appropriate.

Typical LRs for glass evidence lie in the range of 0.1 to 1000. It is difficult to get much below 0.1, which tends to occur when no glass is found but transfer was reasonably likely. Equally, very high LRs, above 1000, happen when a lot of matching glass is found or when two or more controls have matching glass. The scale effectively "runs out of words."

We concede that verbalizing an LR is more difficult than merely "giving a frequency." The very concept of an LR is unfamiliar to most juries, although it seems likely that Bayesian logic is very akin to the natural thinking patterns of most people. These verbal difficulties are part of the barrier to widespread use of LRs. Here, we merely claim that the use of this approach effectively incorporates information about the presence of glass on the clothing per se and that it offers a logical way in which to view glass (and in effect all) evidence. Those persons offering a "frequency" will very shortly get into just as much verbal trouble when they try to either define their frequency or to incorporate data about how rare large groups of glass are on a person's clothing. However, such difficulties are useful in that they encourage discussion and questioning of the evidence and the mode of presentation.

7.2 Sensitivity of the likelihood ratio answer to some of the data estimates

Each of the terms in the LR is an estimate. None of them is known exactly. As such, they have error (sampling or other) in them. Some of these are particularly important in their effect on the LR and some are almost unimportant. In Example 7.1, T_0 is unimportant (it is important whenever a control exists, but there is no matching glass for this control). Again, in this case P_0, P_1, and f are reasonably well supported by data, and while they may have sampling error, this effect is unlikely to be great in comparison to the very difficult terms T_4 and S_4.

When we consider S_4, we find that the difficulty exists in two places. First, S_4 is small, and estimating small things is hard (ask the DNA analysts). The reason for this is quite straightforward. If S_4 is in the order of 0.02 (or less), then we only expect to see it twice in a survey of 100 sets of clothing. Realistically, we might see it a few more times or not at all because of sampling error. It is even more difficult to estimate the higher terms. In a typical case it is not unusual for us to require S_{10} which is very small in the LSH survey for nonmatching glass. Estimating this term with any accuracy is very difficult.

There is one, at least, strong counterargument. Many of the terms are well supported by data. For instance, S_1, S_2, and S_3 appear to be reasonably well estimated. This does not leave much for the other terms (since they must add to 1) so that even if we do not know exactly what they are, we do know that they are small. In addition, the distribution is expected to be continuous; that is, there is no logical reason to expect S_{10} to suddenly be a peak if S_9 and S_{11} are small. In fact, it seems reasonable to expect S_{10} to be smaller than S_9, but larger than S_{11}. This suggests that judgment might be reasonably used to estimate these terms. If S_1 is about 70%, S_2 about 16%, and S_3 about 6%, we have only 8% of the area to distribute among the remaining possibilities. This logic certainly suggests that S_{10} is less than 1%.

These terms could be examined by standard sampling error type techniques, and the most obvious is the bootstrap.[159] The clothing survey (LSH in this case) could be resampled and the relevant terms reestimated many times, producing an estimate of the variance in the LR produced by sampling error. We are unaware of any work done along these lines, and in view of the subjective nature of many of the estimates this may not be warranted.

The second issue arises when there are, say, 100 fresh glass fragments in the debris. The analyst examines ten of these and finds that they all match. Should statements be made about the 10 or the 100? We can certainly guarantee that ten match, but it seems likely that many more do. The difference in the LR equations is quite extreme:

$$LR = T_0 + \frac{P_0 T_{10}}{P_1 S_{10} f}$$

or (7.5)

$$LR = T_0 + \frac{P_0 T_{100}}{P_1 S_{100} f}$$

which clearly would give very different answers. It is safe in this hypothetical case to use

$$LR = T_0 + \frac{P_0 T_{10}}{P_1 S_{10} f}$$

since this is almost certainly the numerically smaller (and hence more conservative) answer.

7.3 The effect of search procedures

Imagine that we had done the previous case differently. We had examined the T-shirt and found the four matching fragments and then stopped. The formula for the LR is still

$$LR = T_0 + \frac{P_0 T_4}{P_1 S_4 f} \quad (7.6)$$

but we need to consider exactly what the *P*, *S*, and *T* terms mean.

P_i is the probability of finding *i* groups of glass under this search procedure. Therefore, in the first example it was under a search of the T-shirt and jeans including pockets. In the second instance it is under a search of the T-shirt only. Glass on the surface of upper clothing is rarer than if the whole set of clothes and the pockets are searched; therefore, we expect the *P* terms to change, reflecting less glass.

S_i is the probability of a group of size *i* under this search procedure.

T_i is the probability that *i* fragments were found using this search procedure given that an arbitrary number were transferred and had persisted.

To demonstrate this effect we take P_0, P_1, and S_4 from the LSH survey of the surfaces of garments only and leave the other terms constant. These are 0.42, 0.26, and 0.01, respectively, giving an LR of 165, which is higher than the 59 achieved by a complete search. Part of this difference is artificial and comes from the loss of information in summarizing the glass as coming from the clothing rather than specifying each position. This was a simplification that we made to facilitate an already complex problem. However, much of the difference is real and suggests that if a large group of matching glass (three or more fragments) is found on the upper clothing or hair, then the search should be stopped.

7.4 Fallacy of the transposed conditional

This term relates to a common logical fallacy also termed the prosecutor's fallacy.[160] This has been well written about[94,102,161] and will only be referred to briefly here. It has been more of an issue in DNA than anywhere else but is, in fact, entirely general.

Let us imagine that the glass evidence warrants an LR of, say, 500 (the frequentist approach is just as susceptible to this fallacy). Strictly, this means that the evidence is 500 times more likely if C is true than if $\bar{C}$ is true. This is a statement about the probability of the evidence, E. The fallacy is to interpret it as a statement about the probability of C (or $\bar{C}$). In words, we (correctly) state that the evidence is 500 times more likely if Mr. Henry broke the window than if he did not. We might erroneously state, or be misconstrued as stating, that it is 500 times more likely that Mr. Henry broke the window.

It is very easy to see this in symbols. The correct expressions say something about $\Pr(E \mid C)$, the probability of the evidence given a hypothesis. The incorrect expressions state something about $\Pr(C)$ or $\Pr(C \mid E)$, the probability of a hypothesis. The name "fallacy of the transposed conditional" comes from equating $\Pr(E \mid C)$ with $\Pr(C \mid E)$. The vertical bar is the conditioning bar, and the terms have been "transposed" about it.

A commonly used analogy goes as follows:

> *The probability that an animal has four legs if it is a cow is very high.*

This statement is not the same as

> *The probability that an animal is a cow if it has four legs is very high.*

The first statement is true and the second is obviously false. This example might seem to suggest that spotting correct and incorrect statements is easy. It is not.

The following are some forensic examples from Evett.[162]

> *The probability of finding this blood type if the stain had come from someone other than Mr. Smith is 1 in 1000.*

This statement is correct. The event is "finding this blood type" and the conditioning information is that it came from some other person. The condition is made clear by the use of "if."

> *The probability that the blood came from someone other than Mr. Smith is 1 in 1000.*

This statement is wrong. It is the most common form of the transposed conditional. It is the spoken equivalent of $Pr(\bar{C} \mid E) = 1/1000$; the probability of a hypothesis given the evidence rather than the other way around.

Both of these statements are reasonably easily assigned as correct or false. However, some are much more difficult:

> *The chance of a man other than Mr. Smith leaving blood of this type is 1 in 1000.*

This statement can be read two ways, and a good discussion can usually ensue.

Guidelines for Avoiding the Transposed Conditional (After Evett)

1. It is inadvisable to speculate on the truth of a hypothesis without considering at least one other hypothesis.
2. Clearly state the alternative hypotheses that are being considered.
3. If a statement is to be made of probability or odds, then it is good practice to use "if" or "given" explicitly to clarify the conditioning information.
4. Do not offer a probability for the truth of a hypothesis.

We drew several conclusions from this discussion. First, it teaches us some guidelines that might keep us from committing this fallacy. These are enumerated below and are well discussed in Evett.[162] However, first it seems worthwhile observing that the frequency with which lawyers, judges, some forensic scientists, and newspapers make this transposition suggests that they want to, understandably, make some statement about C or $\bar{C}$. The observation that such statements are desirable does not in itself show how they can be made. We observe, however, that Bayes Theorem shows in a simple manner how knowledge regarding $Pr(E \mid C)$ can be used to make statements about $Pr(C \mid E)$. We cannot imagine any simpler proof of the need for Bayesian reasoning in the evaluation of evidence. We challenge any person using the frequentist type reasoning to demonstrate any similar simple way in which evidence may be interpreted. Certainly none has been forthcoming.

References

1. Marris, N. A., Interpreting glass splinters, *Analyst*, 59, 686–687, 1934.
2. Nelson, D. F. and Revell, B. C., Backward fragmentation from breaking glass, *J. Forensic Sci. Soc.*, 7, 58–61, 1967.
3. Ojena, S. M. and De Forest, P. R., Precise refractive index determination by the immersion method, using phase contrast microscopy and the mettler hot stage, *J. Forensic Sci. Soc.*, 12, 312–329, 1972.
4. Evett, I. W., The interpretation of refractive index measurements, *Forensic Sci.*, 9, 209–217, 1977.
5. Lindley, D. V., A problem in forensic science, *Biometrika*, 64, 207–213, 1977.
6. Evett, I. W., A Bayesian approach to the problem of interpreting glass evidence in forensic science casework, *J. Forensic Sci. Soc.*, 26, 3–18, 1986.
7. Evett, I. W. and Buckleton, J. S., The interpretation of glass evidence: a practical approach, *J. Forensic Sci. Soc.*, 30, 215–223, 1990.
8. Walsh, K. A. J., Buckleton, J. S., and Triggs, C. M., A practical example of glass interpretation, *Sci. Justice*, 36, 213–218, 1996.
9. Zerwick, C., *A Short History of Glass*, Harry N Adams Publishers, New York, 1990.
10. Tooley, F. V., *The Handbook of Glass Manufacture, Vol. II*, 3rd ed., Ashlee Publishing Co. Inc., New York, 1984.
11. Sanger, D. G., Identification of toughened glass using polarized-light, *J. Forensic Sci. Soc.*, 13, 29–32, 1973.
12. Glathart, J. L. and Preston, F. W., The behaviour of glass under impact: theoretical considerations, *Glass Technol.*, 9, 89–100, 1968.
13. Preston, F. W., A study of the rupture of glass, *J. Soc. Glass Technol.*, 10, 234–269, 1926.
14. Preston, F. W., A study of the rupture of glass, *J. Soc. Glass Technol.*, 10, 234–269, 1926.
15. Murgatroyd, J. B., The significance of surface marks on fractured glass, *J. Soc. Glass Technol.*, 26, 156, 1942.
16. Miller, E. T., Forensic glass comparisons, in *Forensic Science Handbook*, Vol. 1, Saferstein, R., Ed., Prentice-Hall, Englewood Cliffs, NJ, 1982.
17. Oughton, C. D., Analysis of glass fractures, *Glass Ind.*, 26, 72–74, 1945.
18. Renshaw, G. D. and Clarke, P. D. B., Variation in thickness of toughened glass from car windows, *J. Forensic Sci. Soc.*, 14, 311–317, 1974.
19. Buckleton, J. S., Axon, B. W., and Walsh, K. A. J., A survey of refractive index, color and thickness of window glass from vehicles in New Zealand, *Forensic Sci. Int.*, 32, 161–170, 1986.
20. Slater, D. P. and Fong, W., Density, refractive index and dispersion in the examination of glass: their relative worth as proof, *J. Forensic Sci.*, 27, 474–483, 1982.
21. Greene, R. S. and Burd, D. Q., Headlight glass as evidence, *J. Crim. Law Criminol. Police Sci.*, 40, 85, 1949.
22. Fong, W., A thermal apparatus for developing a sensitive, stable and linear density gradient, *J. Forensic Sci. Soc.*, 11, 267–272, 1971.
23. Beveridge, A. D. and Semen, C., Glass density method using a calculating digital density meter, *J. Can. Forensic Sci.*, 12, 113–116, 1979.

24. Gamble, L., Burd, D. Q., and Kirk, P. L., Glass fragments as evidence: a comparative study of physical properties, *J. Crim. Law Criminol.*, 33, 416–421, 1943.
25. Dabbs, M. D. G. and Pearson, E. F., Some physical properties of a large number of window glass specimens, *J. Forensic Sci.*, 17, 70–78, 1972.
26. Stoney, D. A. and Thornton, J. I., The forensic significance of the correlation of density and refractive index in glass evidence, *Forensic Sci. Int.*, 29, 147–157, 1985.
27. Thornton, J. I., Langhauser, C., and Kahane, D., Correlation of glass density and refractive index-implications to density gradient construction, *J. Forensic Sci.*, 29, 711–713, 1984.
28. Pounds, C. A. and Smalldon, K. W. The Efficiency of Searching for Glass on Clothing and the Persistence of Glass on Clothing and Shoes, Forensic Science Service, 1977.
29. Fuller, N. A., Some Observations on the Recovery of Glass by Hair Combing, Metropolitan Police Forensic Science Laboratory, Aldermaston, London, England, 1996.
30. Locke, J. and Scranage, J. K., Breaking of flat glass, part 3: surface particles from windows, *Forensic Sci. Int.*, 57, 73–80, 1992.
31. Elliot, B. R., Goodwin, D. G., Hamer, P. S., Hayes, P. M., Underhill, M., and Locke, J., The microscopical examination of glass surfaces, *J. Forensic Sci. Soc.*, 25, 459–471, 1985.
32. Locke, J. and Zoro, J. A., The examination of glass particles using the interference objective. 1. Methods for rapid sample handling, *Forensic Sci. Int.*, 22, 221–230, 1983.
33. Locke, J. and Elliott, B. R., The examination of glass particles using the interference objective. 2. A survey of flat and curved surfaces, *Forensic Sci. Int.*, 26, 53–66, 1984.
34. Lloyd, J. B. F., Fluorescence spectrometry in the identification and discrimination of float and other surfaces on window glasses, *J. Forensic Sci.*, 26, 325–342, 1981.
35. Locke, J., A simple microscope illuminator for detecting the fluorescence of float glass surfaces, *Microscope*, 35, 151–157, 1987.
36. Zernike, F., Phase contrast, a new method for the microscopic examination of transparent objects, *Physica*, 9, 686–693, 1942.
37. Zoro, J. A., Locke, J., Day, R. S., Bademus, O., and Perryman, A. C., An investigation of refractive index anomalies at the surfaces of glass objects and windows, *Forensic Sci. Int.*, 39, 127–141, 1988.
38. Locke, J., Underhill, M., Russell, P., Cox, P., and Perryman, A. C., The evidential value of dispersion in the examination of glass, *Forensic Sci. Int.*, 32, 217–219, 1986.
39. Cassista, A. R. and Sandercock, P. M. L., Effects of annealing on toughened and non-toughened glass, *J. Can. Soc. Forensic Sci.*, 27, 171–177, 1994.
40. Miller, E. T., A rapid method for the comparison of glass fragments, *J. Forensic Sci.*, 10, 272–281, 1965.
41. Koons, R., Peters, C., and Rebbert, P., Comparison of refractive index, energy dispersive x-ray fluorescence and inductively coupled plasma atomic emission spectrometry for forensic characterization of sheet glass fragments, *J. Anal. Atomic Spectrom.*, 6, 451–456, 1991.

42. Underhill, M., Multiple refractive index in float glass, *J. Forensic Sci. Soc.*, 22, 169–176, 1980.
43. Davies, M. M., Dudley, R. J., and Smalldon, K. W., An investigation of bulk and surface refractive indices of flat window glasses, patterned window glasses, and windscreen glasses, *Forensic Sci. Int.*, 16, 125–137, 1985.
44. Locke, J., Sanger, D. G.;, and Roopnarine, G., The identification of toughened glass by annealing, *Forensic Sci. Int.*, 20, 295–301, 1982.
45. Locke, J. and Hayes, C. A., Refractive index measurements across glass objects and the influence of annealing, *Forensic Sci. Int.*, 26, 147–157, 1984.
46. Underhill, M., The Annealing of Glass. Notes for Caseworkers, Metropolitan Police Forensic Science Laboratory, Aldermaston, London, England, 1992.
47. Locke, J. and Hayes, C. A., Refractive index measurements across a windscreen, *Forensic Sci. Int.*, 26, 147–157, 1984.
48. Locke, J. and Rockett, L. A., The application of annealing to improve the discrimination between glasses, *Forensic Sci. Int.*, 29, 237–245, 1985.
49. Winstanley, R. and Rydeard, C., Concepts of annealing applied to small glass fragments, *Forensic Sci. Int.*, 29, 1–10, 1985.
50. Marcouiller, J. M., A revised glass annealing method to distinguish glass types, *J. Forensic Sci.*, 35, 554–559, 1990.
51. Reeve, V., Mathieson, J., and Fong, W., Elemental analysis by energy dispersive x-ray: a significant factor in the forensic analysis of glass, *J. Forensic Sci.*, 21, 291–306, 1976.
52. Terry, K. W., Van Riessen, A., Lynch, B. F., and Vowles, D. J., Quantitative analysis of glasses used within Australia, *Forensic Sci. Int.*, 25, 19–34, 1984.
53. Ryland, S. G., Sheet or container? — Forensic glass comparisons with an emphasis on source classification, *J. Forensic Sci.*, 31, 1314–1329, 1986.
54. Howden, C. R., Dudley, R. J., and Smalldon, K. W., The analysis of small glass fragments using energy dispersive X-ray fluorescence spectrometry, *J. Forensic Sci. Soc.*, 18, 99–112, 1978.
55. Howden, C. R., The Characterisation of Glass Fragments in Forensic Science with Particular Reference to Trace Element Analysis, University of Strathclyde, Glasgow, 1981.
56. Keeley, R. H. and Christofides, S., Classification of small glass fragments by X-ray microanalysis with the SEM and a small XRF spectrometer, *Scanning Electron Microsc.*, I, 459–464, 1979.
57. Rindby, A. and Nilsson, G., Report on the xrf μ-Beam Spectrometer at the Swedish National Forensic Science Laboratory with Special Emphasis on the Analysis of Glass Fragments, Linköping University, Forensic Science Center, 1995.
58. Coleman, R. F. and Goode, G. C., Comparison of glass fragments by neutron activation analysis, *J. Radioanal. Chem.*, 15, 367–388, 1973.
59. Hickman, D. A., A classification scheme for glass, *Forensic Sci. Int.*, 17, 265–281, 1981.
60. Almirall, J. R., Cole, M., Furton, K. G., and Gettinby, G., Discrimination of glass sources using elemental composition and refractive index: development of predictive models, *Sci. Justice*, 38, 93–101, 1998.
61. Buscaglia, J. A., Elemental analysis of small glass fragments in forensic science, *Anal. Chim. Acta*, 288, 17–24, 1994.

62. Hickman, D. A., Elemental analysis and the discrimination of sheet glass samples, *Forensic Sci. Int.*, 23, 213–223, 1983.
63. Hickman, D. A., Glass types identified by chemical analysis, *Forensic Sci. Int.*, 33, 23–46, 1987.
64. Andrasko, J. and Maehly, A. C., Discrimination between samples of window glass by combining physical and chemical techniques, *J. Forensic Sci.*, 23, 250–262, 1978.
65. Catterick, T. and Hickman, D. A., Sequential multi-element analysis of small fragments of glass by atomic-emission spectrometry using an inductively coupled radiofrequency argon plasma source, *Analyst*, 104, 516–524, 1979.
66. Koons, R., Fiedler, C., and Rawalt, R., Classification and discrimination of sheet and container glasses by inductively coupled plasma-atomic emission spectrometry and pattern recognition, *J. Forensic Sci.*, 33, 49–67, 1988.
67. Wolnik, K., Gaston, C., and Fricke, F., Analysis of glass in product tampering investigations by inductively coupled plasma atomic emission spectrometry with a hydrofluoric acid resistant torch, *J. Anal. Atomic Spectrom.*, 4, 27–31, 1989.
68. Catterick, T. and Hickman, D. A., The quantitative analysis of glass by inductively coupled plasma-atomic emission spectrometry: a five element survey, *Forensic Sci. Int.*, 17, 253–263, 1981.
69. Catterick, T., Hickman, D. A., and Sadler, P. A., Glass glassification and discrimination by elemental analysis, *J. Forensic Sci. Soc.*, 24, 350–350, 1984.
70. Hickman, D. A., Harbottle, G., and Sayre, E. V., The selection of the best elemental variables for the classification of class samples, *Forensic Sci. Int.*, 23, 189–212, 1983.
71. Houk, R. S., Mass spectrometry of inductively coupled plasmas, *Anal. Chem.*, 58, 97A–105A, 1986.
72. Jarvis, K. E., Gray, A. L., and Houk, R. S., *Handbook of Inductively Coupled Plasma Mass Spectrometry*, Blackie & Son Ltd., Glasgow and London, 1992.
73. Zurhaar, A. and Mullings, L., Characterization of forensic glass samples using inductively coupled plasma mass spectrometry, *J. Anal. Atomic Spectrom.*, 5, 611–617, 1990.
74. Parouchais, T., Warner, I. M., Palmer, L. T., and Kobus, H. J., The analysis of small glass fragments using inductively coupled plasma mass spectrometry, *J. Forensic Sci.*, 41, 351–360, 1996.
75. Evett, I. W. and Lambert, J. A., The interpretation of refractive index measurements III, *Forensic Sci. Int.*, 20, 237–245, 1982.
76. Curran, J. M., Triggs, C. M., Buckleton, J. S., and Coulson, S., Combining a continuous Bayesian approach with grouping information, *Forensic Sci. Int.*, 91, 181–196, 1998.
77. Triggs, C. M. and Curran, J. M., A Divisive Approach to the Grouping Problem in Forensic Glass Analysis, University of Auckland, Department of Statistics, Auckland, New Zealand, 1995.
78. Triggs, C. M., Curran, J. M., Buckleton, J. S., and Walsh, K. A. J., The grouping problem in forensic glass analysis: a divisive approach, *Forensic Sci. Int.*, 85, 1–14, 1997.
79. Evett, I. W., The interpretation of refractive index measurements II, *Forensic Sci. Int.*, 12, 37–47, 1978.

80. Scott, A. J. and Knott, M. A., Cluster analysis method for grouping means in the analysis of variance, *Biometrics,* 30, 507–512, 1974.
81. Curran, J. M., Buckleton, J. S., and Triggs, C. M., The robustness of a continuous likelihood approach to Bayesian analysis of forensic glass evidence, *Forensic Sci. Int.,* 104, 93–100, 1999.
82. Welch, B. L., The significance of the difference between two means when the population variances are unequal, *Biometrika,* 29, 350–362, 1937.
83. Hill, G. W., Algorithm 395: student's *t*-distribution, *Commun. ACM,* 13, 617–619, 1970.
84. Best, D. C. and Rayner, J. C. W., Welch's approximate solution for the Behrens-Fisher problem, *Technometrics,* 29, 205–210, 1987.
85. Curran, J. M. and Triggs, C. M., The *t*-test: standard or Welch — which one should we use?, unpublished.
86. Almirall, J. R., Characterization of Glass Evidence by the Statistical Analysis of Their Inductively Coupled Plasma/Atomic Emission Spectroscopy and Refractive Index Data, Nashville, TN, 1996, pp. 1–50.
87. Stoecklein, W., Determination of Source and Characterization of Glass of International Origin, San Antonio, TX, 1996.
88. Johnson, R. A. and Wichern, D. A., *Applied Multivariate Analysis,* Prentice-Hall, Englewood Cliffs, NJ, 1982.
89. Seber, G. A. F., *Multivariate Observations,* John Wiley & Sons, New York, 1984.
90. Stoney, D. A., Evaluation of associative evidence: choosing the relevant question, *J. Forensic Sci. Soc.,* 24, 473–482, 1984.
91. Evett, I. W. and Lambert, J. A., The interpretation of refractive index measurements IV, *Forensic Sci. Int.,* 24, 149–163, 1984.
92. Evett, I. W. and Lambert, J. A., The interpretation of refractive index measurements V, *Forensic Sci. Int.,* 27, 97–110, 1985.
93. Robertson, B. and Vignaux, G. A., Probability — the logic of the law, *Oxford J. Legal Stud.,* 13, 457–478, 1993.
94. Robertson, B. and Vignaux, G. A., *Interpreting Evidence: Evaluating Forensic Science in the Courtroom,* John Wiley & Sons, New York, 1995.
95. Finkelstein, M. O. and Fairley, W. B., A Bayesian approach to identification evidence, *Harv. Law Rev.,* 83, 489–517, 1970.
96. Aitken, C. G. G., *Statistics and the Evaluation of Evidence for Forensic Scientists,* John Wiley & Sons, New York, 1995.
97. Evett, I. W., Bayesian inference and forensic science: problems and perspectives, *Statistician,* 36, 99–105, 1987.
98. Evett, I. W. and Buckleton, J. S., Some aspects of the Bayesian approach to evidence evaluation, *J. Forensic Sci. Soc.,* 29, 317–324, 1989.
99. Taroni, F., Champod, C., and Margot, P. A., Forerunners of Bayesianism in early forensic science, *Jurimetrics,* 38, 183–200, 1998.
100. Bayes, T., An essay towards solving a problem in the doctrine of chances, *Philos. Trans. R. Soc. London for 1763,* 53, 370–418, 1764.
101. Bernardo, J. M. and Smith, A. F. M., *Bayesian Theory,* John Wiley & Sons, New York, 1994.
102. Evett, I. W. and Weir, B. S., *Interpreting DNA Evidence,* Sinauer Associates, Sunderland, MA, 1998.
103. Evett, I. W., Lambert, J. A., and Buckleton, J. S., Further observations on glass evidence interpretation, *Sci. Justice,* 35, 283–289, 1995.

104. Lambert, J. A., Satterthwaite, M. J., and Harrison, P. H., A survey of glass fragments recovered from clothing of persons suspected of involvement in crime, *Sci. Justice*, 35, 273–281, 1995.
105. Harrison, P. H., Lambert, J. A., and Zoro, J. A., A survey of glass fragments recovered from clothing of persons suspecrted of involvement in crime, *Forensic Sci. Int.*, 27, 171–187, 1985.
106. Robertson, B. W. and Vignaux, G. A., DNA evidence: wrong answers or wrong questions, *Genetica*, 96, 145–152, 1995.
107. McQuillan, J. and McCrossan, S., The Frequency of Occurrence of Glass Fragments in Head Hair Samples — a Pilot Investigation, Northern Ireland Forensic Science Laboratory, Belfast, Ireland, 1987.
108. McQuillan, J. and Edgar, K. A., Survey of the distribution of glass on clothing, *J. Forensic Sci. Soc.*, 32, 333–348, 1992.
109. Lindley, D. V., A statistical paradox, *Biometrika*, 44, 187–192, 1957.
110. Pearson, E. F., May, R. W., and Dabbs, M. D. G., Glass and paint fragments found in men's outer clothing, *J. Forensic Sci.*, 16, 283–300, 1971.
111. Davis, R. J. and DeHaan, J. D., A survey of men's footwear, *J. Forensic Sci. Soc.*, 17, 271–285, 1977.
112. Harrison, P. H. The Distribution of Refractive Index of Glass Fragments Found on Shoes, Forensic Science Service, 1978.
113. Hoefler, K., Hermann, P., and Hansen, C. A Study of the Persistence of Glass Fragments on Clothing After Breaking a Window, AN2FSS Symposium, Auckland, New Zealand, 1994.
114. Lau, L., Callowhill, B. C., Conners, N., Foster, K., Gorvers, R. J., Ohashi, K. N., Sumner, A. M., and Wong, H., The frequency of occurrence of paint and glass on the clothing of high school students, *J. Can. Soc. Forensic Sci.*, 30, 233–240, 1997.
115. Zoro, J. A. and Fereday, M. J., Report of a Survey Concerning the Exposure of Individuals to Breaking Glass, Forensic Science Service, Aldermaston, London, England, 1986.
116. Lambert, J. A. and Evett, I. W., The refractive index distribution of control glass samples examined by the forensic science laboratories in the United Kingdom, *Forensic Sci. Int.*, 26, 1–23, 1984.
117. Faulkner, A. E. and Underhill, M., A Collection of Non-Sheet Glass from Non-Casework Sources, Metropolitan Police Forensic Science Laboratory, London, England, 1989.
118. Walsh, K. A. J. and Buckleton, J. S., On the problem of assessing the evidential value of glass fragments embedded in footwear, *J. Forensic Sci. Soc.*, 26, 55–60, 1986.
119. Lambert, J. A. and Satterthwaite, M. J., How likely is it that matching glass will be found in these circumstances, *FSS Report*, 1–15, 1994.
120. Kirk, P. L., *Crime Investigation*, Interscience Publishers, New York, 1953.
121. Locke, J. and Unikowski, J. A., Breaking of flat glass, part 1: size and distribution of particles from plain glass windows, *Forensic Sci. Int.*, 51, 251–262, 1991.
122. Locke, J. and Unikowski, J. A., Breaking of flat glass, part 2: effect of pane parameters on particle distribution, *Forensic Sci. Int.*, 56, 95–102, 1992.
123. Pounds, C. A. and Smalldon, K. W., The distribution of glass fragments in front of a broken window and the transfer of fragments to individuals standing nearby, *J. Forensic Sci. Soc.*, 18, 197–203, 1978.

124. Luce, R. J. W., Buckle, J. L., and McInnis, I. A., Study on the backward fragmentation of window glass and the transfer of glass fragments to individual's clothing, *J. Can. Soc. Forensic Sci.*, 24, 79–89, 1991.
125. Zoro, J. A., Observations on the backward fragmentation of float glass, *Forensic Sci. Int.*, 22, 213–219, 1983.
126. Allen, T. J. and Scranage, J. K., The transfer of glass, part 1: transfer of glass to individuals at different distances, *Forensic Sci. Int.*, 93, 167–174, 1998.
127. Allen, T. J., Hoefler, K., and Rose, S. J., The transfer of glass, part 2: a study of the transfer of glass to a person by various methods, *Forensic Sci. Int.*, 93, 175–193, 1998.
128. Allen, T. J., Hoefler, K., and Rose, S. J., The transfer of glass, part 3: the transfer of glass from a contaminated person to another uncontaminated person during a ride in a car, *Forensic Sci. Int.*, 93, 195–200, 1998.
129. Allen, T. J., Hoefler, K., and Rose, S. J., The transfer of glass, part 4: the transfer of glass fragments from the surface of an item to the person carrying it, *Forensic Sci. Int.*, 93, 201–208, 1998.
130. Francis, K., The Backward Fragmentation Pattern of Glass, as a Result of Impact from a Firearm Projectile, Australian Federal Police, 1993.
131. Locke, J., Badmus, H. O., and Perryman, A. C., A study of the particles generated when toughened windscreens disintegrate, *Forensic Sci. Int.*, 31, 79–85, 1986.
132. Allen, T. J., Locke, J., and Scranage, J. K., Breaking of flat glass, part 4: size and distribution of fragments from vehicle windscreens, *Forensic Sci. Int.*, 93, 209–218, 1998.
133. Hicks, T., Vanina, R., and Margot, P., Transfer and persistence of glass fragments on garments, *Sci. Justice*, 36, 101–107, 1996.
134. Pounds, C. and Smalldon, K., The Distribution of Glass Fragments in Front of a Broken Window and the Transfer of Fragments to Individuals Standing Nearby, Forensic Science Service, 1977.
135. Underhill, M., The acquisition of breaking and broken glass, *Sci. Justice*, 37, 121–129, 1997.
136. Bone, R. G., Secondary Transfer of Glass Fragments Between Two Individuals, Forensic Science Service, Birmingham, England, 1993.
137. Holcroft, G. A. and Shearer, B., The Transfer and Persistence of Glass Fragments with Special Reference to Secondary and Tertiary Transfer, Forensic Science Service, Birmingham, England, 1993.
138. Brewster, F., Thorpe, J. W., Gettinby, G., and Caddy, B., The retention of glass particles on woven fabrics, *J. Forensic Sci.*, 30, 798–805, 1985.
139. Batten, R. A., Results of Window Breaking Experiments in the Birmingham Laboratory, Forensic Science Service, Birmingham, England, 1989.
140. Cox, A. R., Allen, T. J., Barton, S., Messam, P., and Lambert, J. A., The Persistence of Glass. Part 1: the Effects of Cothing Type and Activity, CRSE, 1996.
141. Cox, A. R., Allen, T. J., Barton, S., Messam, P., and Lambert, J. A., The Persistence of Glass. Part 2: the Effects of Fragment Size, Forensic Science Service, Birmingham, England, 1996.
142. Cox, A. R., Allen, T. J., Barton, S., Messam, P., and Lambert, J. A., The Persistence of Glass. Part 3: the Effects of Fragment Shape, Forensic Science Service, Birmingham, England, 1996.

143. Curran, J. M., Triggs, C. M., Hicks, T., Buckleton, J. S., and Walsh, K. A. J., Assessing transfer probabilities in a Bayesian interpretation of forensic glass evidence, *Sci. Justice,* 38, 15–22, 1998.
144. Wright, S., The method of path coefficients, *Ann. Math. Stat.,* 5, 161–215, 1934.
145. Spiegelhalter, D. J., Dawid, A. P., Lauritzen, S. L., and Cowell, R. G., Bayesian analysis in expert systems, *Stat. Sci.,* 8, 219–247, 1993.
146. Spiegelhalter, D. J., Thomas, A., Best, N., and Gilks, W., BUGS — Bayesian Inference Using Gibbs Sampling, MRC Biostatistics Unit, Institute of Public Health, Cambridge, U.K., 1994.
147. Sturges, H. A., The choice of a class interval, *J. Am. Stat. Assoc.,* 21, 65–66, 1926.
148. Hoaglin, D. C., Mosteller, F., and Tukey, J. W., *Understanding Robust and Exploratory Data Analysis,* John Wiley & Sons, New York, 1983.
149. Dixon, W. J. and Kronmal, R. A., The choice of origin and scale for graphs, *J. Assoc. Comput. Mach.,* 12, 259–261, 1965.
150. Scott, D. W., On optimal and data-based histograms, *Biometrika,* 66, 605–610, 1979.
151. Freedman, D. and Diaconis, P., On the maximum deviation between the histogram and and the underlying density, *Z. Wahrscheinlichkeitstheorie verwandte Gebiete,* 58, 139–167, 1981.
152. Freedman, D. and Diaconis, P., On the histogram as a density estimator: L2 theory, *Z. Wahrscheinlichkeitstheorie verwandte Gebiete,* 57, 453–456, 1981.
153. Scott, D. W., *Multivariate Density Estimation,* John Wiley & Sons, New York, 1992.
154. Curran, J. M., Triggs, C. M., Almirall, J. R., Buckleton, J. S., and Walsh, K. A. J., The interpretation of elemental composition measurements from forensic glass evidence, *Sci. Justice,* 37, 241–244, 1997.
155. Curran, J. M., Triggs, C. M., Almirall, J. R., Buckleton, J. S., and Walsh, K. A. J., The interpretation of elemental composition measurements from forensic glass evidence II, *Sci. Justice,* 37, 245–249, 1997.
156. Buckleton, J. S. and Walsh, K. A. J., *The Use of statistics in Forensic Science,* Aitken, C. G. G. and Stoney, D. A., Eds., Ellis Horwood Ltd., Chichester, 1991.
157. Taroni, F. and Aitken, C. G. G., Forensic science at trial, *Jurimetrics,* 37, 327–337, 1997.
158. Evett, I. W., Towards a uniform framework for reporting opinions in forensic science casework, *Sci. Justice,* 38, 198–202, 1998.
159. Efron, B., Bootstrap methods: another look at the jackknife, *Ann. Stat.,* 7, 1–26, 1979.
160. Thompson, W. C. and Schumann, E. L., Interpretation of statistical evidence in criminal trials: the prosecutor's fallacy and the defence attorney's fallacy, *Law Hum. Behav.,* 11, 167–187, 1987.
161. Aitken, C. G. G. and Stoney, D. A., *The Use of Statistics in Forensic Science,* Ellis Horwood Ltd., Chichester, 1991.
162. Evett, I. W., Avoiding the transposed conditional, *Sci. Justice,* 35, 127–131, 1995.

Index

D

E

F

G

H

I

J

K

L

M

N

O

P

Q

R

S

T

U

V

W

X